# Ergebnisse der Mathematik und ihrer Grenzgebiete

Band 59

Herausgegeben von
P. R. Halmos · P. J. Hilton · R. Remmert · B. Szőkefalvi-Nagy

Unter Mitwirkung von
L. V. Ahlfors · R. Baer · F. L. Bauer · R. Courant · A. Dold
J. L. Doob · S. Eilenberg · M. Kneser · G. H. Müller · M. M. Postnikov
B. Segre · E. Sperner

Geschäftsführender Herausgeber: P. J. Hilton

S. López de Medrano

# Involutions on Manifolds

Springer-Verlag New York Heidelberg Berlin 1971

Dr. Santiago López de Medrano
Instituto de Matemáticas
Universidad Nacional Autónoma de México

AMS Subject Classifications (1970): 57-02, 57 D XX, 57 E XX

ISBN-13:978-3-642-65014-7 e-ISBN-13:978-3-642-65012-3
DOI: 10.1007/978-3-642-65012-3


Softcover reprint of the hardcover 1st edition 1971

S. López de Medrano

# Involutions on Manifolds

Springer-Verlag Berlin Heidelberg New York 1971

Dr. Santiago López de Medrano
Instituto de Matemáticas
Universidad Nacional Autónoma de México

AMS Subject Classifications (1970): 57-02, 57 D XX, 57 E XX

ISBN-13:978-3-642-65014-7 e-ISBN-13:978-3-642-65012-3
DOI: 10.1007/978-3-642-65012-3

*A mi madre*

*In Memoriam*
*Dr. Nicanor Almarza y Herranz*
*(1898–1968)*

# Table of Contents

# Introduction

This book contains the results of work done during the years 1967–1970 on fixed-point-free involutions on manifolds, and is an enlarged version of the author's doctoral dissertation [54] written under the direction of Professor William Browder.

The subject of fixed-point-free involutions, as part of the subject of group actions on manifolds, has been an important source of problems, examples and ideas in topology for the last four decades, and receives renewed attention every time a new technical development suggests new questions and methods ([62, 8, 24, 63]). Here we consider mainly those properties of fixed-point-free involutions that can be best studied using the techniques of surgery on manifolds. This approach to the subject was initiated by Browder and Livesay. Special attention is given here to involutions of homotopy spheres, but even for this particular case, a more general theory is very useful. Two important related topics that we do not touch here are those of involutions with fixed points, and the relationship between fixed-point-free involutions and free $S^1$-actions. For these topics, the reader is referred to [23], and to [33], [61], [82], respectively.

The two main problems we attack are those of classification of involutions, and the existence and uniqueness of invariant submanifolds with certain properties. As will be seen, these problems are closely related.

If $(T, \Sigma^n)$ is a fixed-point-free involution of a homotopy sphere $\Sigma^n$, the quotient $\Sigma^n/T$ is called a homotopy projective space. This terminology is justified in IV.3.1 where it is shown that there is a homotopy equivalence between $\Sigma^n/T$ and real projective space $P^n$. Furthermore this homotopy equivalence is unique (up to homotopy) if we specify that it be orientation-preserving when $n$ is odd.

Therefore, to classify fixed-point-free involutions of homotopy spheres is equivalent to classifying manifolds with the same homotopy type as $P^n$. In order to classify manifolds within a given homotopy type it is convenient to follow what we may call the Browder-Novikov-Sullivan program:

(i) Classify the manifolds up to normal cobordism (III.1).

(ii) Within each normal cobordism class (normal invariant) classify up to addition of an element $\theta^n(\partial\pi)$.

(iii) Determine the action of $\theta^n(\partial\pi)$ on these manifolds.

In the case of simply-connected manifolds, Browder and Novikov gave the answer to (ii): within each normal cobordism class there is only one manifold, modulo the action of $\theta^n(\partial\pi)$ ($n \geqq 5$), and there is a nice obstruction theory for deciding when a normal invariant (III.1) is the normal cobordism class of a manifold within the given homotopy type. Then Sullivan showed how to reduce (i) to homotopy theory in a very natural way, and solved the homotopy problem corresponding to the p.l. case, thus giving an effective procedure for computing the set of normal invariants.

The homotopy projective spaces, not being simply-connected, represent the simplest case where the result of Browder and Novikov no longer holds as stated above. For them, and for other manifolds with fundamental group $\mathbb{Z}_2$, we show (IV.3.3 and V.2) that within each normal cobordism class there are, modulo the action of $\theta^n(\partial\pi)$:

a) only one manifold if $n \neq 4k+3$, $n \geqq 5$, and

b) infinitely many, classified by an invariant $\sigma \in \mathbb{Z}$ (the Browder-Livesay invariant) when $n = 4k+3$, $k \geqq 1$.

In the p.l. case we are able to carry out step (i), using results of Sullivan and Wall, and some geometric arguments, while step (iii) is trivial, so we obtain the

**Classification Theorem.** For $k>0$:

$$hT(P^{4k+1}) \approx \mathbb{Z}_4 \oplus (2k-2)\,\mathbb{Z}_2$$

$$hT(P^{4k+2}) \approx \mathbb{Z}_4 \oplus (2k-2)\,\mathbb{Z}_2$$

$$hT(P^{4k+3}) \approx \mathbb{Z}_4 \oplus (2k-2)\,\mathbb{Z}_2 \oplus \mathbb{Z}$$

$$hT(P^{4k+4}) \approx \mathbb{Z}_4 \oplus (2k-1)\,\mathbb{Z}_2.$$

Here $k\mathbb{Z}_2$ denotes the direct sum $\mathbb{Z}_2 \oplus \mathbb{Z}_2 \oplus \cdots \oplus \mathbb{Z}_2$, ($k$ times).

The finite part gives the normal invariant and the infinite cyclic part for $hT(P^{4k+3})$ gives the Browder-Livesay invariant.

The set $hT(P^n)$ denotes the set of p.l. equivalence classes of p.l. involutions of $S^n$, or equivalently, the set of p.l. homeomorphism classes of p.l. manifolds homotopy equivalent to $P^n$.

For the smooth case, step (i) presents considerable homotopy theoretical difficulties, for to compute the set of normal cobordism classes we would have to know not only the homotopy groups of $G/0$

(or, equivalently, the stable homotopy groups of spheres) but also have some knowledge of its $k$-invariants. We treat here only the involutions of homotopy 7-spheres. For completeness we include a theorem of Browder about the action of $\theta^{4k+3}(\partial\pi)$, that solves step (iii) in a good number of cases.

Wall obtained a different version of the classification theorem which uses an invariant $\tau \in \mathbb{Z}$ instead of the Browder-Livesay invariant $\sigma$. The expected result, $\sigma = \tau$, turned out surprisingly to be a consequence of simultaneous work by Hirzebruch, based on the Atiyah-Bott-Singer fixed point theorem. We give here a proof of this result (up to a sign), and derive many interesting consequences. We also give the sketch of a proof of the result, without the sign indeterminacy.

The other main problem we consider is the existence of invariant spheres for involutions of homotopy spheres. The existence and uniqueness of codimension 1 spheres was solved theoretically by Browder and Livesay who defined invariants (one of which is the $\sigma$ mentioned above) that give the complete obstruction to these problems. We complete their results with the

**Realization Theorem.** *In dimensions* $\geqq 5$, *all the Browder-Livesay obstructions for the existence and uniqueness of codimension 1 spheres can be realized.*

The precise statement is given on II.1. The detailed study of the examples and their connections gives additional information.

We also consider the existence and uniqueness problem for invariant codimension 2 spheres, and invariant codimension 2, unknotted spheres, and we give a complete answer. For invariant spheres of higher codimension, the problems can best be attacked with the various known embedding and isotopy theorems.

As corollaries we get some non-embedding and non-isotopy results, examples of involutions of homology 3-spheres, applications to Brieskorn involutions, etc.

The above problems admit generalizations to involutions of simply-connected manifolds, and we give also some results on the realization of Browder-Livesay invariants in this situation.

In the presentation of these results we have preferred to introduce the simplest cases first (for which more detailed information can usually be obtained) and then proceed gradually to the more general cases.

Chapter 0 gives a summary of the notation, conventions and background material we will use.

Chapter I contains a description of the results of Browder and Livesay, the definition of the invariants and the proof of some elementary properties of them.

Chapter II starts with the statement of the realization theorem. The proof of its several parts is scattered through Chapters II to V. After some elementary constructions with involutions are described, the chapter continues with the proof of some parts of the realization theorem, and we point out some consequences, such as a non-isotopy result, the existence of invariant codimension 2 spheres in some cases (a subject that will be treated more fully in Chapter VI), the construction of involutions of homology 3-spheres and counterexamples to a generalization of a result of Siebenmann.

In Chapter III we give a brief review of the theory of normal invariants, including Sullivan's version, and of Wall's theory of non-simply-connected surgery. Then we establish the relation between Wall's obstructions and the Browder-Livesay invariants, and using this we prove another part of the realization theorem and a lemma that was used in Chapter II.

Chapter IV is dedicated to the proof of the classification theorem and starts by a computation of the set of p.l. normal invariants of $P^n$. We then prove two surgery theorems, that can be better proved in the general situation of involutions of simply-connected manifolds. Putting all our previous information together, we obtain the classification theorem, which in turn solves another case of the realization theorem. Then we give our proof of the $\sigma=(\pm)\tau$ result and draw some consequences. Finally we mention briefly the topological classification of homotopy projective spaces, which follows from the work of Kirby and Siebenmann.

Chapter V starts with some general remarks on the difficulties that arise in the classification of smooth involutions, and we give some upper bounds on their number. Then we give a more precise version of part of the realization theorem, and we generalize it to obtain the realization of some Browder-Livesay invariants for involutions of simply-connected manifolds. Using this, the $\sigma=(\pm)\tau$ result extends easily to this more general situation and we obtain several consequences, one of which is the elimination of an artificial connectivity hypothesis in the generalized Browder-Livesay theorem. As another consequence we obtain a relation between the differentiable structures of the manifolds and the Browder-Livesay invariants of the involutions within a normal invariant. Then we prove the remaining case of the realization theorem by first giving explicit examples on Brieskorn spheres and, in the following section, by a more illuminating construction due to Browder. To describe this construction we quote the necessary definitions and results about the generalized Kervaire invariant. As a by-product we show that $\theta^n_I$, the set of homotopy $n$-spheres that admit fixed-point-free involutions, is a subgroup of $\theta^n$. We then compute the smooth involutions of 7-spheres and the subgroup $\theta_I^n$ for $n \leqq 15$, and state another result of Browder on the action of $\theta^{4k+3}(\partial\pi)$ on homotopy projective spaces. The chapter ends with the sketch of a proof of the $\sigma = \tau$ result.

In Chapter VI we study codimension 2 invariant spheres. We discuss the differences between invariant and characteristic spheres and prove that the difference disappears when the codimension is not bigger than 2. We apply this to prove a non-embedding theorem, to study some involutions of Brieskorn spheres, and to obtain existence and uniqueness theorems for codimension 2 knotted and unknotted invariant spheres. The realization theorem gives directly the realization of the obstructions that arise in this case, and we can obtain several examples and a non-isotopy result. In the last section of the book we start the study of the knot-cobordism class of an invariant $S^{4k+1}$ for $(T, \Sigma^{4k+3})$.

We have not tried to push our results to their greatest generality, such as involutions of non-simply-connected manifolds (though we mention how one of our constructions can be generalized: V.2.2), involutions of more general systems of manifolds, or "involutions" of Poincaré spaces. Our many relations between Browder-Livesay invariants and abstract surgery obstructions suggest that a general theory of Browder-Livesay invariants would be very fruitful. For example, out of our arguments of Chapters IV and VI it can be easily proved the fact that $L_{4k+3}(\mathbb{Z}_2, +) \approx L_{4k}(\mathbb{Z}_2, -) \approx \mathbb{Z}_2$, by proving Theorem 1, IV.3.3, without having to appeal to these computations. Other obstruction groups for $\mathbb{Z}_2$ have been computed geometrically by Wall [81] using the Browder-Livesay approach.

I want to express here my deep gratitude to my thesis advisor, Professor William Browder, who introduced me to the problems discussed here, gave me the right suggestions at the right time and listened patiently to my chaotic expositions and helped weed out the nonsense.

For very helpful conversations and explanations, thanks are due to George Cooke, Francisco González Acuña, Larry Siebenmann, Dennis Sullivan and C.T.C. Wall. I wish to thank also Professors J. Cerf, F. Hirzebruch, G.R. Livesay, D. Montgomery and P. Mostert for their kind invitations to participate in various meetings and activities.

I wish to thank also all my friends and teachers, both in Mexico and Princeton, who for many years helped me and encouraged me in uncountably many ways. For this, it is an honor to be indebted to Professors Solomon Lefschetz and Norman E. Steenrod.

During the preparation of this work I was financially supported by the Instituto de Matemáticas, UNAM, and the Instituto Nacional de la Investigación Científica. For their efforts on my behalf, thanks are due to Professors José Adem, Alfonso Nápoles Gándara and Roberto

Vázquez García. I also received partial support from Princeton University and a travel grant from the Centro de Formación de Profesores e Investigadores, UNAM.

Finally, I would like to mention the name of a dear late friend. Dr. Nicanor Almarza was an outstanding scientist and committed teacher whose work was interrupted by war and political exile. A tragic accident prevented him from seeing this work completed, just a month before these lines were written. To his warm friendship and to his positive and understanding influence, which I have been fortunate to enjoy for as long as I can remember, the following pages should be attributed.

Chapter 0

# Notation, Conventions, Preliminaries

Throughout this work $(T, M^n)$ will denote a fixed point free involution $T: M^n \to M^n$ of a compact manifold $M$ of dimension $n$, with or without boundary. Only in Section V.8 will we consider involutions with fixed points, and we will state carefully which involutions are fixed point free and which are not. We will always assume that $M$ is a combinatorial (piecewise linear, p.l.) manifold and that $T$ is piecewise linear (p.l.), or that $M$ is a smooth ($C^\infty$) manifold and that $T$ is smooth. Then the quotient $M/T$ gets a well defined piecewise linear or smooth structure.

We will assume known the basic results and constructions of combinatorial and differential topology (as exposed, for example, in [84] and [24]) such as transverse regularity ($t$-regularity), general position, straightening of angles, normal bundles, tubular and collar neighborhoods, etc., and we will use frequently their equivariant versions, which can be thought of in terms of the quotient manifolds. For example, we say that an equivariant map $f: (T, M) \to (T', M')$ is in general position if the induced map $f/T: M/T \to M'/T'$ is.

In the expository sections we will talk briefly about Poincaré complexes, Poincaré pairs, and maps of degree 1 between them ([11, 18, 73, 78, 80, 82]).

All manifolds, Poincaré complexes, etc., will be assumed *oriented* whenever they are orientable, and all homotopy equivalences, p.l. homeomorphisms, diffeomorphisms, etc., will be assumed orientation-preserving, with the important exception of the involutions themselves which can be orientation-reversing. When $(T, M)$ is orientation preserving we will always assume that the quotient is oriented in such a way that the projection $M \to M/T$ preserves the orientation locally. If $M$ is an oriented manifold we will assume that $\partial M$ is oriented by $[\partial M] = \partial_* [M]$ and when we say that $W$ is an oriented cobordism between $M_0$ and $M_1$ it should be understood that $\partial_* [W] = [M_1] - [M_0]$.

We say that an equivariant map $f: (T, M) \to (T', M')$ is an equivariant homotopy equivalence if $f: M \to M'$ is a homotopy equivalence. We say that $f$ is an equivalence between the involutions $(T, M)$ and

$(T', M')$ if $f$ is a p.l. homeomorphism or a diffeomorphism, depending on the category we are considering.

$\Sigma^n$ will denote a homotopy sphere, while $S^n$ will denote the standard one. Therefore $(T, \Sigma^n)$ will denote a fixed point free involution of a homotopy sphere, and the quotient will be called a homotopy projective space (IV.3.1). The antipodal map of $S^n$ will be denoted by $(a, S^n)$ and its quotient is the standard projective space $P^n$. To denote a homotopy projective space we usually use the symbol $Q^n$.

In this and other situations where no confusion can arise we will pass unconsciously from the involution to the quotient and back to the double covering.

$\theta^n$ is the group of $h$-cobordism classes of homotopy spheres and $\theta^n(\partial\pi) = bP^{n+1}$ the subgroup of those that bound $\pi$-manifolds ([42]). These groups act on smooth manifolds by connected sum (see, for example, [13]).

A symmetric quadratic form $(G, A, \phi)$ over the finitely generated free abelian group $G$ consists of a bilinear map $A\colon G \times G \to \mathbb{Z}$ and a quadratic map $\phi\colon G \to \mathbb{Z}$ associated with $A$, i.e., $\phi(kx) = k^2\phi(x)$ and $\phi(x+y) = \phi(x) + \phi(y) + A(x, y)$. This implies that $A$ is even (i.e. $A(x, x)$ is even for all $x \in G$) and symmetric $(A(x, y) = A(y, x))$ and $A$ and $\phi$ determine each other by the above formula and $\phi(x) = \frac{1}{2}A(x, x)$. We say that $(G, A, \phi)$ is unimodular if $A$ is (i.e. if $x \to A(x, \ )$ gives an isomorphism $G \to \mathrm{Hom}(G, \mathbb{Z})$, or, equivalently, if the matrix of $A$ with respect to some basis of $G$ has determinant $\pm 1$) and in that case the index of $A$ is a multiple of 8 ([57, 35]) and we define $\sigma(G, A, \phi) = \frac{1}{8}$ (Index $A$). The value of $\sigma$ determines $(G, A, \phi)$ up to a stable equivalence ([42, 58, 70]).

A skew-symmetric quadratic form $(G, A, \phi)$ over $G$ consists of a bilinear map $A\colon G \times G \to \mathbb{Z}$ and a quadratic map $\phi\colon G \to \mathbb{Z}_2$ associated with $A$, i.e., $\phi(kx) = k^2\phi(x)$ and $\phi(x+y) = \phi(x) + \phi(y) + A(x, y)$, mod. 2, such that $A(x, y) = -A(y, x)$. $\phi$ determines $A$ mod. 2, but different $\phi$'s can be associated to the same $A$. We say that $(G, A, \phi)$ is unimodular if $A$ is, and in that case the Arf invariant $c(G, A, \phi) = c(\phi) \in \mathbb{Z}_2$ is defined ([2, 16, 39]). The value of $c$ and the rank of $G$ determine $(G, A, \phi)$ up to an equivalence. The same results apply when $G$ is a finitely generated $\mathbb{Z}_2$-vector space and $A\colon G \times G \to \mathbb{Z}_2$.

If $M^{2k}$ is a $(k-1)$-connected framed manifold whose boundary is a homotopy sphere, we can associate to it a unimodular quadratic form $(G, A, \phi)$, symmetric if $k$ is even, skew-symmetric if $k$ is odd, where $G = H_k(M)$, $A(x, y) = x \cdot y$ (intersection number) and $\phi(x) = \frac{1}{2}(x \cdot x)$ if $k$ is even and $\phi$ is the quadratic map defined in [39, 42] if $k$ is odd, and $\sigma(M) = \sigma(G, A, \phi) = \frac{1}{8}$ Index $M$ and $c(M) = c(G, A, \phi)$ respectively. The Milnor manifold $M_0^{4k}$ is the manifold of this type with rank $H_{2k}(M_0) = 8$ and $\sigma(M_0^{4k}) = 1$ for $k > 1$, or any manifold of this type with $\sigma(M_0^4) = 2$, rank $H_2(M_0^4)$ unspecified and $\partial M_0^4 = S^3$, for $k=1$. For $k>1$, $\partial M_0^{4k} = \Sigma_0^{4k-1}$ generates $\theta^{4k-1}(\partial\pi)$ and the closed Milnor manifold $\bar{M}_0^{4k}$ is obtained from $M_0^{4k}$ by attaching a cone on its boundary. The Kervaire manifold $K^{4k+2}$ is the manifold of this type with rank $H_{2k+1}(K^{4k+2})=2$ and $c(K^{4k+2})=1$. $\partial K^{4k+2} = \Sigma_0^{4k+1}$ generates $\theta^{4k+1}(\partial\pi)$ and the closed Kervaire manifold $\bar{K}^{4k+2}$ is obtained from $K^{4k+2}$ by attaching a cone on its boundary ([42]).

We will assume known the theory of simply connected surgery ([11, 18, 19, 42, 58, 65, 78]) and in particular the definition of the surgery obstructions: If $X^{4k}$ is an orientable Poincaré complex and $f\colon M^{4k} \to X^{4k}$ is a normal map (III.1) then $\sigma(f) = \frac{1}{8}(\mathrm{Index}(M) - \mathrm{Index}(X))$ and can also be defined in terms of the symmetric intersection form defined on $\ker(H_{2k}(M) \to H_{2k}(X))$. The notation $\sigma(f)$ will be used also for the Browder-Livesay invariant of an equivariant homotopy equivalence $f$, I.2.2, but this should not cause any confusion. If $f\colon M^{4k+2} \to X^{4k+2}$ is a normal map then $c(f) \in \mathbb{Z}_2$ is defined by making $f$ $(2k+1)$-connected and then defining the skew-symmetric quadratic form $(G, A, \phi)$ on $G = \ker(H_{2k+1}(M) \to H_{2k+1}(X))$ as in [16].

Since we will be using a great number of maps and the kernels of the induced maps in homology, we will use the notation

$$K_i(X \to Y) = \ker(H_i(X) \to H_i(Y)).$$

Again the notation $K_i(f)$ will have a special meaning (I.2.2) which should not cause confusion.

We will make extensive use of the $h$-cobordism theorem ([59, 72]) and of its equivariant version, which holds because of the $s$-cobordism theorem ([40]) and the fact that $Wh(\mathbb{Z}_2)=0$ ([30]).

Chapter I

# The Browder-Livesay Invariants

## I.1 Involutions of Spheres

### I.1.1 Desuspensions and Characteristic Submanifolds

**Definition.** We say that an involution $(T, \Sigma^n)$ of a homotopy sphere $\Sigma^n$ *desuspends* if there is an embedded (smoothly or p.l., depending on the category we're working on) homotopy sphere $S^{n-1} \subset \Sigma^n$ which is invariant under $T$. We also say that $(T, \Sigma^n)$ is a *suspension* of $(T|S^{n-1}, S^{n-1})$ and that $(T|S^{n-1}, S^{n-1})$ is a *desuspension* of $(T, \Sigma^n)$. See also IV.3.2 and V.1.

We describe here a procedure due to Browder and Livesay for deciding whether an involution $(T, \Sigma^n)$ desuspends or not. The idea is to start with an invariant submanifold $W^{n-1} \subset \Sigma^n$ and to try to simplify it by equivariant surgery until one gets a homotopy sphere. Since an invariant sphere must divide $\Sigma^n$ in two parts, and the involution must interchange them (otherwise there would at least be one fixed point in each), we must start with an invariant submanifold that has this property.

**Definition.** If $W^{n-1} \subset \Sigma^n$ is a submanifold such that $W = A \cap TA$ and $\Sigma = A \cup TA$, where $A$ is a compact submanifold of $\Sigma$ with $\partial A = W$, then we say that $(T|W, W)$ is a *characteristic submanifold for* $(T, \Sigma^n)$. We will also call the quotient $W/T$ a characteristic submanifold for the homotopy projective space $\Sigma/T$, and, loosely speaking, call $W$ itself a characteristic submanifold for $(T, \Sigma^n)$.

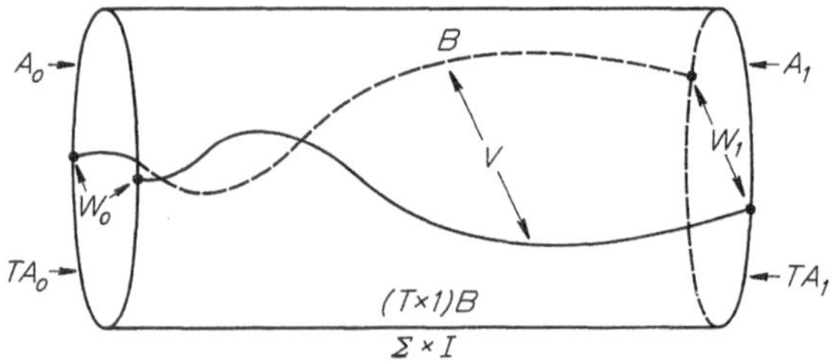


Fig. 1

If $W_0^{n-1}$ and $W_1^{n-1}$ are two characteristic submanifolds for $(T, \Sigma^n)$, a *characteristic cobordism* between $W_0$ and $W_1$ is an involution $(T\times 1|V, V)$, where $V$ is a submanifold of $\Sigma\times I$ that intersects $\Sigma\times\{i\}$ transversely in $W_i\times\{i\}$, $i=0, 1$, such that $V=B\cap(T\times 1)B$, $\Sigma\times I=B\cup(T\times 1)B$, where $B$ is a compact submanifold of $\Sigma\times I$ with boundary $A_0\times\{0\}\cup V\cup A_1\times\{1\}$ (and corners along $W_i\times\{i\}$).

**Lemma.** *For every* $(T, \Sigma)$ *there is a characteristic submanifold, and given two characteristic submanifolds, there is a characteristic cobordism joining them.*

*Proof.* The double covering $p\colon \Sigma\to\Sigma/T$ has a classifying map $g\colon \Sigma/T\to P^N$, $N$ large, which we can assume $t$-regular at $P^{N-1}$. Then $p^{-1}g^{-1}(P^{N-1})$ is a characteristic submanifold for $(T, \Sigma)$. Since any characteristic submanifold can be obtained in this way, the second assertion follows from the relative $t$-regularity theorem.

### I.1.2 Equivariant Surgery

If $W=A\cap TA$ is a characteristic submanifold for $(T, \Sigma^n)$, let

$$K_q=K_q(W)=K_q(W\to A).$$

It follows easily from the Mayer-Vietoris sequence of the triad $(\Sigma; A, TA)$ that

$$H_q(W)=K_q\oplus T_*(K_q).$$

Therefore if $W$ is simply connected and $K_q(W)=0$ for $q\leqq[n/2]$, then $H_q(W)=0$ for $q\leqq n-2$ and $W$ is the homotopy sphere we're looking for.

If $f\colon (D^{q+1}, S^q)\to(A, W)$ is an embedding such that $f(D^{q+1}-S^q)\subset A-W$, $f(D^{q+1})$ intersects $W$ transversely and $f(S^q)\cap Tf(S^q)=\phi$, then we can find a tubular neighborhood $N$ of $(f(D^{q+1}), f(S^q))$ in $(A, W)$ such that $N\cap TN=\phi$. If we form $A'=\overline{A-N\cup T(N)}$ (with straightened corners),

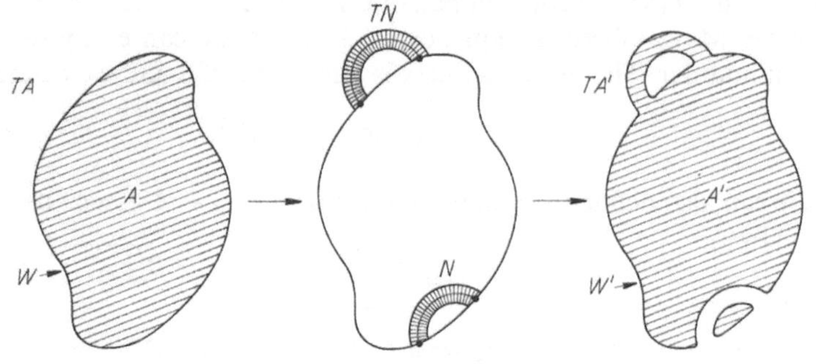


Fig. 2

$W' = A' \cap TA'$ is a new characteristic submanifold for $(T, \Sigma)$. We say that $W'$ has been obtained from $W$ by performing equivariant surgery at $f$.

**Lemma.** *By a sequence of equivariant surgeries we can make* $W$ $[(n-3)/2]$*-connected.*

*Proof.* Assuming $W$ is $(q-1)$-connected, $q \leqq [(n-3)/2]$ and given $\alpha \in K_q(W)$, by general position we can find an embedding $f: (D^{q+1}, S^q) \to (A, W)$ such that $f(S^q)$ represents $\alpha$ and which we can use to perform equivariant surgery, obtaining a characteristic submanifold $W'$ with $K_q(W') = K_q(W)/(\alpha)$. After a finite number of such surgeries we obtain a $q$-connected characteristic submanifold. To start the induction we use the same argument to kill $\ker(\pi_0(W) \to \pi_0(A))$ and $\ker(\pi_1(W) \to \pi_1(A))$.

### I.1.3 The Desuspension Theorem

To every involution $(T, \Sigma^n)$ we associate an invariant $\sigma(T, \Sigma^n)$ called the Browder-Livesay invariant of $(T, \Sigma^n)$. It lies in the following groups

$$\sigma(T, \Sigma^n) \in \begin{cases} 0 & n \text{ even} \\ \mathbb{Z} & n \equiv 3 \text{ mod. } 4 \\ \mathbb{Z}_2 & n \equiv 1 \text{ mod. } 4. \end{cases}$$

It is defined as follows: consider an $[(n-3)/2]$-connected characteristic submanifold for $(T, \Sigma^n)$.

If $n = 4k+3$ we have the bilinear form

$$B(x, y) = x \cdot T_* y$$

defined on $K_{2k+1}$. Since the intersection form is skew-symmetric and $T$ reverses the orientation, $B(x, y)$ is symmetric. If we represent $x \in K_{2k+1}$ by an embedded sphere $S^{2k+1}$ in general position, the intersection points of $S^{2k+1}$ and $T(S^{2k+1})$ come in pairs $(p, Tp)$. Therefore the form is even. From the Mayer-Vietoris sequence of $(\Sigma, A, TA)$ we can easily see that it is unimodular. Therefore its index is a multiple of 8 and we define

$$\sigma(T, \Sigma^{4k+3}) = \tfrac{1}{8}(\text{Index } B(x, y))^1.$$

We can also define the quadratic form $\psi: K_{2k+1} \to \mathbb{Z}$ associated with $B(x, y)$ by

$$\psi(x) = \frac{x \cdot T_* x}{2}.$$

---

[1] This differs from the definition given in [21], [22], [52] and [53], where the factor $\frac{1}{8}$ was not included.

If $n=4k+1$, the form $B(x, y)=x \cdot T_* y$ on $K_{2k}$ is now skew-symmetric and unimodular. Again, the intersection of a sphere in general position representing $x \in K_{2k}$, with its image under $T$ consists of pairs of points and letting $\psi(x)$ be the number of such pairs mod. 2 we have defined

$$\psi: K_{2k} \to \mathbb{Z}_2$$

which is quadratic with respect to $B(x, y)$, i.e.

$$\psi(x+y)=\psi(x)+\psi(y)+x \cdot T_* y \qquad \text{mod. } 2$$

and the Arf invariant $c(\psi) \in \mathbb{Z}_2$ is defined. Set

$$\sigma(T, \Sigma^{4k+1})=c(\psi).$$

If we consider two $[(n-3)/2]$-connected characteristic submanifolds for $(T, \Sigma^n)$, by Lemma I.1.1 there is a characteristic cobordism joining them, and the standard argument applies to show that $\sigma$ is well defined (cf. I.2, Lemma 1).

**Theorem** (Browder and Livesay [21, 22]). *If* $n \geqq 6$, $(T, \Sigma^n)$ *desuspends if, and only if,* $\sigma(T, \Sigma^n)=0$.

*Sketch of proof.* If $(T, \Sigma^n)$ desuspends, then the invariant sphere serves as a characteristic submanifold for $(T, \Sigma^n)$ and obviously $\sigma(T, \Sigma^n)=0$ without any conditions on $n$.

If $n=2k$, we can always, by general position, represent elements in $K_{k-1}$ by embedded spheres disjoint from their images, but by performing equivariant surgery we don't necessarily reduce $K_{k-1}$. An adaptation of the method described in [42] shows that the sequence of surgeries can be chosen in such a way that we finally kill all of $K_{k-1}$. Therefore $(T, \Sigma^{2k})$ always desuspends.

If $n$ is odd, $\sigma(T, \Sigma^n)$ is precisely the obstruction to finding a basis $e_1, \ldots, e_r, f_1, \ldots, f_r$ for $K_{(n-1)/2}$ in terms of which the quadratic form is given by

$$B(e_i, f_j)=\delta_{ij},$$
$$B(e_i, e_j)=B(f_i, f_j)=0,$$
$$\psi(e_i)=\psi(f_i)=0.$$

Once we have such a basis, by getting rid of the pairs of intersection points, two at a time, we can find embedded disjoint spheres representing the elements of this basis, disjoint from their images, and such that the actual number of intersection points of one with the image of another one is the algebraic intersection number. Then a sequence of surgeries on the spheres representing $e_1, \ldots, e_r$ reduces their number while preserving these properties and at the end we arrive at a homotopy sphere.

### I.1.4 Concordance of Desuspensions

**Definition.** Two desuspensions $(T|S_i^{n-1}, S_i^{n-1})$, $i=0, 1$, of $(T, \Sigma^n)$ are *concordant* if there is a characteristic cobordism joining them which is an $h$-cobordism.

For $n \geqq 6$, if two desuspensions are concordant then they are equivalent as involutions, by the $s$-cobordism theorem. As it will be seen in the next theorem and in IV.4.2, the converse is true for $n \neq 4k+2$, $(n \neq 3)$. The case $n=4k+2$ is discussed in II.3.

The process described above can be applied to this situation: perform equivariant surgery on a characteristic cobordism joining $S_0^{n-1}$ and $S_1^{n-1}$, with a concordance as objective. As before we can define the Browder-Livesay invariant $\sigma(T, \Sigma^n, S_0^{n-1}, S_1^{n-1})$, which measures the obstruction to completing this surgery process, and lies in the following groups:

$$\sigma(T, \Sigma^n, S_0^{n-1}, S_1^{n-1}) \in \begin{cases} 0 & n \text{ odd} \\ \mathbb{Z} & n \equiv 0 \text{ mod. } 4 \\ \mathbb{Z}_2 & n \equiv 2 \text{ mod. } 4. \end{cases}$$

**Theorem** (Browder and Livesay [21, 22]). *Let $(T|S_i^{n-1}, S_i^{n-1})$, $i=0, 1$, be two desuspensions for $(T, \Sigma^n)$, $n \geqq 5$. Then $(T|S_0^{n-1}, S_0^{n-1})$ and $(T|S_1^{n-1}, S_1^{n-1})$ are concordant if, and only if, $\sigma(T, \Sigma^n, S_0^{n-1}, S_1^{n-1})=0$.*

## I.2 Involutions of Simply Connected Manifolds

### I.2.1 Characteristic Submanifolds and $h$-regularity

**Definition.** Let $(T, M^n, \partial M)$ be a fixed point free involution of a manifold $M$. A *characteristic submanifold* is an embedded submanifold $(W^{n-1}, \partial W) \subset (M, \partial M)$ such that

a) $\partial M = A \cup TA$, $\partial W = A \cap TA$, where $A$ is a compact submanifold of $\partial M$ with boundary $\partial W$,

b) $M = B \cup TB$, $W = B \cap TB$, where $B$ is a compact submanifold of $M$ with boundary $A \cup W$ (and corner along $\partial W$).

As in I.1, there is always a characteristic submanifold for $(T, M, \partial M)$, any characteristic submanifold for $(T|\partial M, \partial M)$ is the boundary of one for $(T, M, \partial M)$, and similar results hold for characteristic cobordisms, etc.

For simplicity we will assume from now on that $\partial M = \phi$.

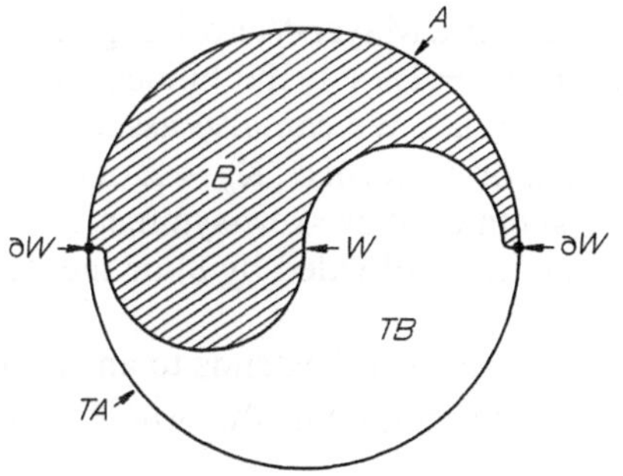


Fig. 3

Let $f$: $(T_1, M_1) \to (T_0, M_0)$ be an equivariant homotopy equivalence and $W_0$ a characteristic submanifold for $(T_0, M_0)$.

**Definition.** $f$ is $h$-regular at $W_0$ if

a) $f$ is $t$-regular at $W_0$, and therefore $f^{-1}(W_0) = W_1$ is a characteristic submanifold for $(T_1, M_1)$,

b) $f|W_1$: $W_1 \to W_0$ is a homotopy equivalence.

### I.2.2 The Browder-Livesay Invariants

Let $BL_n(\varepsilon)$, where $\varepsilon = +$ or $-$ be the groups given by the following table

| $n$ (mod. 4) | 0 | 1 | 2 | 3 |
|---|---|---|---|---|
| $BL_n(+)$ | 0 | $\mathbb{Z}_2$ | 0 | $\mathbb{Z}$ |
| $BL_n(-)$ | 0 | $\mathbb{Z}$ | 0 | $\mathbb{Z}_2$ |

**Definition.** Let $f$: $(T_1, M_1) \to (T_0, M_0)$ be an equivariant homotopy equivalence, where $\pi_1(M_0) = 0$. The *Browder-Livesay invariant of* $f$ is an element

$$\sigma(f) \in BL_n(\varepsilon)$$

where $n = \dim M_0$ and $\varepsilon = +$ or $-$ according to whether $T$ is orientation preserving or not, and it is defined as follows:

Let $W_0$ be a characteristic submanifold for $(T_0, M_0)$. We can assume, by performing equivariant surgery if necessary, that $\pi_1(W_0) = 0$. Make $f$ $t$-regular at $W_0$, $W_1 = f^{-1}(W_0)$. Then $W_0 = A_0 \cap T_0 A_0$, $W_1 = A_1 \cap T_1 A_1$, and $f^{-1}(A_0) = A_1$. We can assume again that $\pi_1(W_1) = 0$, and it follows that $\pi_1(A_0) = \pi_1(A_1) = 0$. Let

$$K_i = K_i(f, W_1, W_0) = K_i(W_1 \to W_0) \cap K_i(W_1 \to A_1).$$

As before, it follows from a Mayer-Vietoris argument that $K_i(W_1 \to W_0) = K_i \oplus T_{1*} K_i$, and when $n=2m+1$, the bilinear form $B(x,y)=x \cdot T_{1*} y$ defined on $K_m$ is unimodular. We can also assume that $K_i=0$ for $i<m$ and then we have the quadratic form $\psi: K_m \to \mathbb{Z}$ or $\psi: K_m \to \mathbb{Z}_2$, according to whether $B(x,y)$ is symmetric or skew-symmetric, which is defined as in I.1.3, and $\sigma(f)$ is the index (divided by 8) or the Arf invariant of this quadratic form.

We say that $f$ is a *boundary* if $f$ extends to an equivariant homotopy equivalence $F:(T_1', N_1, \partial N_1) \to (T_0', N_0, \partial N_0)$, where $(T_i' | \partial N_i, \partial N_i) = (T_i, M_i)$, $i=0,1$, and $\pi_1(N_0)=0$.

**Lemma 1.** *If $f$ is a boundary, then $\sigma(f)=0$.*

*Proof.* Let $V_0 = B_0 \cap T_0' B_0$ be a characteristic submanifold for $(T_0', N_0)$ whose boundary is $W_0$, and $V_1 = F^{-1}(V_0)$, $B_1 = F^{-1}(B_0)$. We can assume again that $\pi_1(V_i)=\pi_1(B_i)=0$. Let $K_i(F, V_1, V_0) = K_i(V_1 \to V_0) \cap K_i(V_1 \to B_1)$, then $K_i(V_1 \to V_0) = K_i(F, V_1, V_0) \oplus T_{1*} K_i(F, V_1, V_0)$ and if $n=2m+1$, $L = \ker(K_m(f, W_1, W_0) \to K_m(F, V_1, V_0))$ is a subgroup of half the rank on which $B(x,y)$ is identically 0. If we are in one of the cases where $B(x,y)$ is skew-symmetric, and $K_i(f, W_1, W_0)=0$ for $i<m$, we can assume $K_i(F, V_1, V_0)=0$ for $i<m$. Then we can represent $x \in L$ as the boundary of an embedded disc $(D^{m+1}, S^m) \subset (V_1, W_1)$, in general position with respect to $T_1(D^{m+1})$. $D^{m+1} \cap T_1(D^{m+1})$ consists of a collection of arcs $\gamma_i$ whose boundaries form $S^m \cap T_1(S^m)$, and a collection of circles. Since $T_1(\gamma_i)$ is another arc of the collection, and is different from $\gamma_i$ ($T_1$ fixed point free), the number of such arcs is even and the number of pairs of intersection points in $S^m \cap T_1(S^m)$ is also even, i.e. $\psi(x)=0$. Therefore $\psi$ is identically 0 on $L$ and $\sigma(f)=0$.

**Corollary.** *$\sigma(f)$ depends only on the equivariant homotopy class of $f$, i.e., is independent of the choices of $W_0$ and $W_1$.*

**Theorem** (Browder and Livesay, unpublished). *Let $f$: $(T_1, M_1^n) \to (T_0, M_0^n)$, $n \geqq 6$ be an equivariant homotopy equivalence and $W_0 = A_0 \cap T_0 A_0$ a characteristic submanifold for $(T_0, M_0)$. Assume $M_0$ and $W_0$ are 1-connected and, if $n$ is odd, that the pair $(A_0, W_0)$ is 2-connected. Then $f$ is equivariantly homotopic to a map h-regular at $W_0$ if, and only if, $\sigma(f)=0$.*

*Remarks.* This theorem generalizes I.1.3, and further generalizations can be made, like the analog of I.1.4, absolute and relative versions for manifolds with boundary and more general systems of manifolds.

In V.2.3, we give a proof of some cases of this theorem with the connectivity condition for $(A_0, W_0)$ removed.

We need a further property of $\sigma$. Let $f_2$: $(T_2, M_2) \to (T_1, M_1)$ and $f_1: (T_1, M_1) \to (T_0, M_0)$ be equivariant homotopy equivalences, $\pi_1(M_i)=0$, and assume we're in one of the cases where $\sigma \in \mathbb{Z}$. Let dim $M_0 = 2m+1$.

**Lemma 2.** $\sigma(f_1 f_2)=\sigma(f_1)+\sigma(f_2)$.

*Proof.* Let $W_0$ be a characteristic submanifold for $(T_0, M_0)$. Assume $f_1$ $t$-regular at $W_0$, $f_1^{-1}(W_0)=W_1$, $f_2$ $t$-regular at $W_1$, $f_2^{-1}(W_1)=W_2$. Then we have the following commutative diagram and split exact sequences

$$\begin{array}{ccccccc} 0\to K_*(W_2\to W_1) & \longrightarrow & H_*(W_2) & \underset{\alpha}{\rightleftarrows} & H_*(W_1) & \to 0 \\ \downarrow & & \downarrow & & \downarrow & \\ 0\to K_*(A_2\to A_1) & \longrightarrow & H_*(A_2) & \underset{\beta}{\rightleftarrows} & H_*(A_1) & \to 0 \end{array}$$

where $\alpha=P(f_2|W_2)^* P^{-1}$, $\beta=P(f_2|A_2)^* P^{-1}$ and $P$ represents the Poincaré duality maps. Then, $\alpha$ preserves intersection numbers and $K_*(W_2\to W_1)$ and $\alpha H_* W_1$ are orthogonal under intersections. Therefore

$$H_m(W_2)=K_m(W_2\to W_1)\oplus H_m(W_1)$$

gives an orthogonal direct sum decomposition of the form $x\cdot y$. Intersecting with $K_m(W_2\to W_0)$ we get

$$K_m(W_2\to W_0)=K_m(W_2\to W_1)\oplus K_m(W_1\to W_0)$$

and intersecting with $K_m(W_2\to A_2)$ and using the commutativity of the above diagram, we finally get:

$$K_m(f_1 f_2, W_2, W_0)=K_m(f_2, W_2, W_1)\oplus K_m(f_1, W_1, W_0)$$

which gives an orthogonal direct sum decomposition of the form $B(x, y)$, and the lemma follows.

*Remark.* An *intrinsic* Browder-Livesay invariant can be defined for an involution $(T, M^n)$ for $n=4k+3$, $T$ orientation preserving, or $n=4k+1$, $T$ orientation reversing. $\sigma(T, M)$ is the index (divided by 8) of the form $x\cdot T_* y$ defined on

$$K_{(n-1)/2}(N\to A)/\mathrm{im}\,\Delta$$

where $N=A\cap TA$ is a characteristic submanifold for $(T, M)$ and $\Delta\colon H_{(n+1)/2}(M)\to H_{(n-1)/2}(N)$ is the Mayer-Vietoris boundary.

Then for an equivariant homotopy equivalence $f\colon (T_1, M_1)\to (T_2, M_2)$ we have

$$\sigma(f)=\sigma(T_1, M_1)-\sigma(T_2, M_2).$$

So $\sigma(f)$ depends only on $(T_1, M_1)$ and $(T_2, M_2)$ and not on the equivariant homotopy equivalence.

Chapter II

# Realization of the Browder-Livesay Invariants

## II.1 The Realization Theorem

**Theorem.** A) *For every* $k>0$ *and every* $i \in \mathbb{Z}$ *there is an involution* $(T, \Sigma^{4k+3})$ *such that* $\sigma(T, \Sigma^{4k+3})=i$.

B) *For every* $k>0$ *there is an involution* $(T, \Sigma^{4k+1})$ *such that* $\sigma(T, \Sigma^{4k+1})=1$.

C) *For every* $k>0$ *and every* $i \in \mathbb{Z}$, *given an involution* $(T, \Sigma^{4k+4})$ *and an invariant* $S_0^{4k+3} \subset \Sigma^{4k+4}$, *there is an invariant* $S_1^{4k+3} \subset \Sigma^{4k+4}$ *such that* $\sigma(T, \Sigma^{4k+4}, S_0^{4k+3}, S_1^{4k+3})=i$.

D) *For every* $k>0$, *given an involution* $(T, \Sigma^{4k+2})$ *and an invariant* $S_0^{4k+1} \subset \Sigma^{4k+2}$, *there is an invariant* $S_1^{4k+1} \subset \Sigma^{4k+2}$ *such that* $\sigma(T, \Sigma^{4k+2}, S_0^{4k+1}, S_1^{4k+1})=1$.

*Commentaries.* Part A is due to the author [52], part B is due to C.T.C. Wall [83] and the author [53] in the p.l. case, and to I. Bernstein [5] and W. Browder [17] in the smooth case. Several independent proofs of parts C and D have been given ([6, 54, 83]).

The proofs of these results will be given in this and the following chapters (A: II.4, B: IV.3.5 (p.l.) and V.4 (smooth), C: III.4 and D: II.3). Some improvements and refinements of parts of this theorem are given with the proofs and also in IV.3.4, IV.4.2 and in Chapter V. In Chapter V the realization of Browder-Livesay invariants in the more general situation of I.2 is considered.

## II.2 Constructions with Involutions

### II.2.1 $M \cup_T M^*$

Given a manifold $M^n$ with boundary $\partial M = A^{n-1} \cup B^{n-1}$, $A \cap B = \partial A = \partial B$, and an involution $T\colon A \to A$ we can form another manifold $M'^n$ and an involution $T'\colon M' \to M'$ as follows:

Consider another copy $M^*$ of $M$, and if $x \in M$, denote by $x^*$ the corresponding point of $M^*$. Let $M' = M \cup_{(T,A)} M^*$, that is $M'$ is obtained from the disjoint union of $M$ and $M^*$ by identifying $x^*$ and $T(x)$ for all $x \in A$. Define $T'$: $M' \to M'$ by $T'(x) = x^*$ and $T'(x^*) = x$, which is compatible with the identifications because $T$ is an involution.

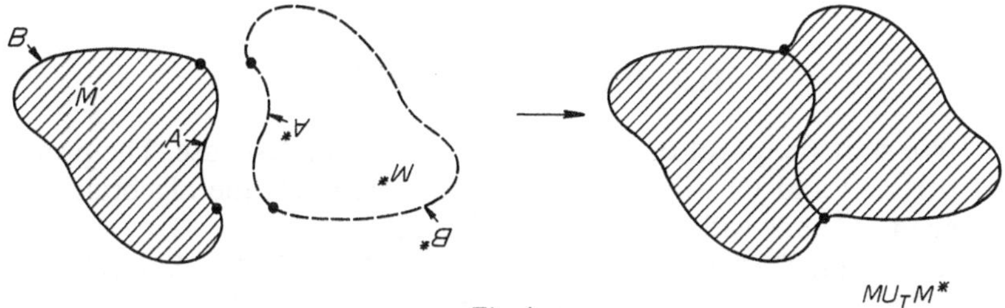


Fig. 4

Then: a) $M'$ is a manifold with boundary $B' = B \cup_{(T, \partial B)} B^*$ (corners must be straightened in the smooth case).

b) $T'$ is an involution of $M'$ and the inclusion $(A, \partial A) \subset (M', \partial M')$ is equivariant.

c) The fixed points of $T'$ are those of $T$ in $A$, so $T'$ is fixed point free if, and only if, $T$ is. If $T$ is fixed point free, then $(T, A, \partial A)$ is a characteristic submanifold for $(T', M', \partial M')$. In fact, any involution $(T, N)$ is of the form $M \cup_{(T,A)} M^*$, where $A = M \cap TM$ is a characteristic submanifold for $(T, N)$.

d) The quotient $M'/T'$ can be obtained from $M$ by identifying $x$ with $T(x)$ for all $x \in A$.

An important special case of this construction is the suspension (see IV.3.2 and V.1).

## II.2.2 $(T, M^n) \# N$

Given a fixed point free involution $(T, M^n)$ and a manifold $N^n$, we can form the equivariant connected sum as follows. Let $p$ be a point in the interior of $M$, $D_1^n$ a small disc around $p$, $p^* = T(p)$, $D_1^{n*} = T(D_1^n)$, and we can assume $D_1^n \cap D_1^{n*} = \phi$. Take two copies $N$ and $N^*$ of $N$ and $D_2^n$ an embedded disc in the interior of $N$, $D_2^{n*}$ the corresponding disc in $N^*$. If $M, N$ are oriented and $T$ is orientation preserving, one should choose the embedding $D_1 \subset M$ orientation preserving and the embedding $D_2 \subset N$ orientation reversing. Let $M'$ be obtained from $M - (D_1 \cup D_1^*) \cup (N - D_2) \cup (N^* - D_2^*)$ by identifying $\partial D_1 = \partial D_2$ and $\partial D_1^* = \partial D_2^*$, and $T'$: $M' \to M'$ given by $T'(x) = T(x)$ for $x \in M - (D_1 \cup D_1^*)$,

$T'(x)=x^*$ for $x\in N-D_2$, $T'(x^*)=x$ for $x^*\in N^*-D_2^*$. $(T', M')$ is a fixed point free involution, which will be denoted by $(T, M)\# N$. The quotient $M'/T'$ is the ordinary connected sum $M/T\# N$.

Similarly we can define connected sum along the boundary by taking $D_1^n$ and $D_2^n$ to be half-disc neighborhoods of points in $\partial M$ and $\partial N$, respectively.

## II.3 Proof of Theorem II.1-D

We start the proof of Theorem II.1 by its simplest case. The proofs of parts A) and C) will be of a similar nature, but the algebraic and geometric details will be more complicated.

Given $(T, \Sigma^{4k+2})$ and an invariant $S_0^{4k+1}\subset\Sigma^{4k+2}$, $S_0=D_0^{4k+2}\cap TD_0^{4k+2}$, we want to show that there exists $S_1^{4k+1}\subset\Sigma$ such that $\sigma(T, \Sigma, S_0, S_1)=1$.

Consider $S_0\times I$ with the involution $T_1=T|S_0\times 1$. Let $K^{4k+2}$ be the Kervaire manifold and $(T_2, W)=(T_1, S_0\times I)\# K$, the equivariant connected sum along the boundary $S_0\times\{1\}$ (II.2.2). Then $H_{2k+1}(W)$ is free on four generators $e_1, f_1, e_2, f_2$ with $e_i\cdot f_j=\delta_{ij}$, $e_i\cdot e_j=f_i\cdot f_j=0$, $\phi(e_i)=\phi(f_i)=1$, and $T_{2*}(e_1)=e_2$, $T_{2*}(f_1)=f_2$. Consider the new basis

$$e_1'=e_1+f_2, \qquad f_1'=e_1+f_1+f_2,$$
$$e_2'=e_2+f_1+f_2, \qquad f_2'=-e_2-f_1.$$

This is again a symplectic basis: $e_i'\cdot f_j'=\delta_{ij}$, $e_i'\cdot e_j'=f_i'\cdot f_j'=0$ and now $\phi(e_i')=\phi(f_i')=0$. Also $T_{2*}(e_1')=-f_2'$, $T_{2*}(e_2')=f_1'$. Therefore we can perform surgery on $e_1'$ and $e_2'$, to kill $H_{2k+1}(W)$. The trace of these surgeries is a cobordism between $W$ and an $h$-cobordism which we can identify with $S_0\times I$, and attach $D_0\times I$ to it to obtain a manifold $V$. Then $\partial V=D_0\times\{0\}\cup W\cup D_0\times\{1\}$, $\pi_1(V)=0$, $H_i(V)=0$ for $i\neq 2k+1$, $H_{2k+1}(V)$ is free on two generators $a_1$, $a_2$ and for the inclusion map $c\colon W\to V$ we have $c_*(e_i')=0$, $c_*(f_i')=a_i$. Now apply the construction II.2.1 to obtain a manifold $N=V\cup_{(T_2, W)}V^*$, with a fixed point free involution, such that $\pi_1(N)=0$, $H_0(N)=H_{4k+2}(N)=\mathbb{Z}$ and $H_i(N)=0$ for $i\neq 0$, $4k+2$. This follows from the Mayer-Vietoris sequence of $(N; V, V^*)$, since the only non-trivial part is

$$\begin{array}{c} 0\to H_{2k+2}(N)\to H_{2k+1}(W)\to H_{2k+1}(V)\oplus H_{2k+1}(V^*)\to H_{2k+1}(N)\to 0 \\ \searrow^{(c_*,\, c_* T_{2*})} \qquad \wr\wr \\ H_{2k+1}(V)\oplus H_{2k+1}(V) \end{array}$$

The triangle is commutative and the map $(c_*, c_* T_{2*})$ is given by $e_1'\to(0, -a_2)$, $e_2'\to(0, a_1)$, $f_i'\to(a_i, 0)$ and is therefore an isomorphism.

By the $s$-cobordism theorem, we can identify equivariantly $N$ with $\Sigma \times I$ and $W$ is a characteristic cobordism between $S_0$ and $S_1^{4k+1} = W \cap \Sigma \times \{1\}$. $K_{2k+1} = K_{2k+1}(W \to V)$ is free on $e_1', e_2'$ and for these we have

$$e_1' \cdot T_{2*} e_2' = e_1' \cdot f_1' = 1$$
$$\psi(e_1') = \psi(e_1) + \psi(f_2) + e_1 \cdot T_{2*} f_2 = 1$$
$$\psi(e_2') = \psi(e_2) + \psi(f_1 + f_2) + e_2 \cdot T_{2*}(f_1 + f_2) = 1$$

($\psi(e_i) = \psi(f_i) = 0$ since we can represent them by spheres disjoint from their images under $T_2$). Therefore

$$\sigma(T, \Sigma, S_0, S_1) = c(\psi) = \psi(e_1') \cdot \psi(e_2') = 1.$$

The proof also yields the following corollaries: Since the obstruction to concordance lies in $\mathbb{Z}_2$, any other invariant $S^{4k+1} \subset \Sigma^{4k+2}$ is concordant to either $S_0^{4k+1}$ or $S_1^{4k+1}$. Now $(T|S_1, S_1)$ is from the construction equivalent to the connected sum $(T|S_0, S_0) \# \Sigma_0^{4k+1}$, where $\Sigma_0^{4k+1} = \partial K$ is the Kervaire sphere, and we get

**Corollary 1.** *In the p.l. case, any two desuspensions of* $(T, \Sigma^{4k+2})$ *are equivalent as involutions, though not always concordant.*

**Corollary 2.** *There are two concordance classes of p.l. embeddings of* $P^{4k+1}$ *in* $P^{4k+2}$ *for* $k>0$. *In particular, there are two non-isotopic embeddings.*

This last corollary was first noticed by Berstein and Livesay ([6]). For $n \not\equiv 1 \pmod{4}$, $n \geq 5$, any two embeddings of $P^n$ in $P^{n+1}$ are concordant. See also VI.3.

These corollaries hold also in the smooth case in the dimensions where the Kervaire sphere is standard. This is known to be the case for $k = 1, 3, 7, 15$ ([16]).

## II.4 Proof of Theorem II.1-A

Given $(T, \Sigma^{4k+3})$, $k>0$, consider any $2k$-connected characteristic submanifold $W = A \cap TA$. Let $\alpha_1, \dots, \alpha_m$ be a basis for $K_{2k+1} = K_{2k+1}(W \to A)$. Then $\alpha_1, \dots, \alpha_m, T_* \alpha_1, \dots, T_* \alpha_m$ is a basis for $H_{2k+1}(W) = K_{2k+1} \oplus T_* K_{2k+1}$. Let $H$ be the matrix $(\alpha_i \cdot T_* \alpha_j)$.

*Definition.* A set of elements $\alpha_1', \dots, \alpha_m'$ of $H_{2k+1}(W)$ is *admissible* if

(i) $\alpha_i' \cdot \alpha_j' = 0$.

(ii) $\phi(\alpha_i') = 0$.

(iii) The matrix $(\alpha_i' \cdot T_* \alpha_j')$ is unimodular.

Notice that the matrix $\alpha_i' \cdot T_* \alpha_j'$ is always even and symmetric (see I.1.3).

Clearly $\alpha_1, \ldots, \alpha_m$ is an admissible set.

**Lemma 1.** *A set of elements $\alpha_1', \ldots, \alpha_m' \in H_{2k+1}(W)$ is admissible if, and only if, there is a fixed point free involution $(T', \Sigma'^{4k+3})$ with characteristic submanifold $(T, W)$ for which $\alpha_1', \ldots, \alpha_m'$ form a basis of $K_{2k+1}' = K_{2k+1}(W \to A')$.*

*Proof.* One of the implications was noted above. Assume $\alpha_1', \ldots, \alpha_m'$ is admissible. Then $\alpha_1', \ldots, \alpha_m', T_* \alpha_1', \ldots, T_* \alpha_m'$ is a basis for $H_{2k+1}(W)$ since the matrix of intersection numbers is (by (ii))

$$\begin{pmatrix} 0 & H' \\ -H' & 0 \end{pmatrix}$$

where $H' = (\alpha_i' \cdot T_* \alpha_j')$, and is therefore unimodular by condition (iii). Now conditions (i) and (ii) imply that we can perform framed surgery on $\alpha_1', \ldots, \alpha_m'$. The trace of this surgery is a framed cobordism between $W$ and a homotopy sphere, this sphere is framed cobordant to 0, $(W = \partial A)$ and is therefore diffeomorphic to $S^{4k+2}$. Attaching a disc to this sphere[2] we obtain a manifold $A'$ with boundary $W$. $A'$ is $2k$-connected and $H_{2k+1}(A')$ is free on $m$ generators $a_1, \ldots, a_m$ such that the inclusion $c_*: H_{2k+1}(W) \to H_{2k+1}(A')$ is given by $\alpha_i' \to 0$, $T_* \alpha_i' \to a_i$. Now apply construction II.2.1 to obtain a manifold $\Sigma'^{4k+3} = A' \cup_{(T,W)} A'^*$ with an involution $T'$. $\pi_1(\Sigma') = 0$ and $\Sigma'$ is $(4k+2)$-connected as follows from the Mayer-Vietoris sequence of $(\Sigma', A', A'^*)$, the only non-trivial part being

$$\begin{array}{ccccccccc} 0 \to H_{2k+2}(\Sigma') \to H_{2k+1}(W) \to H_{2k+1}(A') \oplus H_{2k+1}(A'^*) \to H_{2k+1}(\Sigma') \to 0 \\ \qquad\qquad {\scriptstyle (c_*,\, c_* T_*)} \searrow \qquad\qquad \wr\wr \qquad\qquad \\ \qquad\qquad\qquad\qquad H_{2k+1}(A') \oplus H_{2k+1}(A') \end{array}$$

and the map $(c_*, c_* T_*)$ is given by $\alpha_i' \to (0, a_i)$, $T_* \alpha_i' \to (a_i, 0)$ and is therefore an isomorphism. So $\Sigma'$ is a homotopy sphere, $(T, W)$ is a characteristic submanifold for $(T', \Sigma')$ and $\alpha_1', \ldots, \alpha_m'$ is a basis of $K_{2k+1}'$ as required.

Now we investigate the algebraic conditions corresponding to those in the definition of an admissible basis. Write

$$\alpha_i' = p_{ij} \alpha_j + q_{ij} T_* \alpha_j \quad \text{(summation convention)}.$$

---

[2] There are many ways of doing it. Cf. V.2.

Then

$$\alpha_i' \cdot \alpha_j' = p_{ik} q_{jl} h_{kl} - q_{ik} p_{jl} h_{kl}$$

$$\begin{aligned}\phi(\alpha_i') &= \phi(p_{ij}\alpha_j) + \phi(q_{ij} T_* \alpha_j) + (p_{ij}\alpha_j) \cdot (q_{ik} T_* \alpha_k) \\ &= 0 + 0 + \sum_j p_{ij} h_{jk} q_{ik}\end{aligned}$$

$$\alpha_i' \cdot T_* \alpha_j' = p_{ik} p_{jl} h_{kl} - q_{ik} q_{jl} h_{kl}.$$

Where $H=(h_{ij})=(\alpha_i \cdot T_* \alpha_j)$ and $P=(p_{ij})$ and $Q=(q_{ij})$ are two $m \times m$ matrices. So conditions (i), (ii) and (iii) of the definition of admissible basis can be restated as

(i)′ $PHQ^t$ is symmetric,

(ii)′ $PHQ^t$ is even,

(iii)′ $H' = PHP^t - QHQ^t$ is unimodular.

*Definition.* Let $H$ and $H'$ be two unimodular, even, symmetric matrices of the same rank. We say that $H'$ is *compatible* with $H$ if there are square matrices $P$ and $Q$ such that $PHQ^t$ is even and symmetric and $H' = PHP^t - QHQ^t$.

Then Lemma 1 can be restated as

**Lemma 1′.** *$H'$ is compatible with $H$ if, and only if, there is a fixed point free involution $(T', \Sigma'^{4k+3})$ with characteristic submanifold $(T, W)$ for which the form $x \cdot T_* y$ in $K'_{2k+1}$ has matrix $H'$ with respect to some basis.*

A few lines of computations can show that compatibility is an equivalence relation, but in fact we have:

**Lemma 2.** *Any two unimodular, even, symmetric matrices of the same rank are compatible.*

This lemma will be proved in III.2.3. A direct proof can be given of the fact that any two such matrices are "stably" compatible, that is, that $H \oplus U_m$ is compatible with $H' \oplus U_m$ for some $m$, where $U_m$ denotes the $2m \times 2m$ matrix with 1's in the non-principal diagonal and 0's elsewhere. For this one checks directly that the matrices

$$P = \begin{pmatrix} 2&1&2&1&0&0&1&3 \\ 2&1&2&1&0&1&1&2 \\ 1&1&1&1&0&1&2&1 \\ 1&1&1&1&0&0&1&1 \\ 1&0&1&1&1&1&0&1 \\ 0&0&2&1&1&1&0&1 \\ 0&0&1&1&1&0&0&0 \\ 1&0&0&1&1&0&0&0 \end{pmatrix}, \quad Q = \begin{pmatrix} 2&1&1&1&0&1&1&2 \\ 1&1&2&1&0&1&1&3 \\ 1&1&2&1&0&0&1&2 \\ 1&0&1&1&0&0&2&1 \\ 1&0&2&1&1&0&1&1 \\ 1&0&1&1&1&1&0&0 \\ 0&0&0&1&0&1&0&0 \\ 0&0&1&0&1&0&0&1 \end{pmatrix}$$

give the compatibility relation between $U_4$ and the well-known matrix

$$E_8=\begin{pmatrix} 2&1&0&0&0&0&0&0\\ 1&2&1&0&0&0&0&0\\ 0&1&2&1&0&0&0&0\\ 0&0&1&2&1&0&0&0\\ 0&0&0&1&2&1&0&1\\ 0&0&0&0&1&2&1&0\\ 0&0&0&0&0&1&2&0\\ 0&0&0&0&1&0&0&2 \end{pmatrix}$$

which has rank 8 and index 8 $(\sigma(E_8)=1)$. Since any unimodular, even, symmetric matrix is stably equivalent to the direct sum of a number of copies of $E_8$ and $U_m$, [70], the result follows. This weaker result is enough for the applications, by adding more handles, if necessary, to the characteristic submanifolds involved. The explicit matrices $P$ and $Q$ will be used in IV.4. Notice that these differ slightly from the ones given in [52], [53], because the conditions they are asked to satisfy have been changed to adapt them to the more general situation considered here.

Using Lemma 2, (or the weaker result stated above), we are in a position to prove the following strong form of Theorem II. 1-A:

**Theorem.** *Given $(T, \Sigma^{4k+3})$, $k>0$, and $i \in \mathbb{Z}$, there is a $2k$-connected characteristic submanifold $(T, W)$ for $(T, \Sigma)$ and an involution $(T', \Sigma'^{4k+3})$ with the same characteristic submanifold $(T, W)$ such that $\sigma(T', \Sigma')=i$.*

*Proof.* Let $W_1$ be a $2k$-connected characteristic submanifold for $(T, \Sigma)$ and $H'$ a non-singular, even, symmetric matrix with $\sigma(H')=i$ and rank $H' \geqq \frac{1}{2}$ rank $H_{2k+1}(W_1)$. By adding handles equivariantly to $W_1$ inside $\Sigma$ we can obtain a new characteristic submanifold $W$ such that rank $H' = \frac{1}{2}$ rank $H_{2k+1}(W)$ (rank $H'$ and $\frac{1}{2}$ rank $H_{2k+1}(W_1)$ are both even numbers). Then the matrix $H$ of the form $x \cdot T_* y$ with respect to some basis of $K_{2k+1}$ has the same rank as $H'$, by Lemma 2, $H'$ is compatible with $H$, and the theorem follows from Lemma 1'.

In V.3 we describe the relation between the homotopy spheres $\Sigma$ and $\Sigma'$.

## II.5 Homology 3-Spheres

For $k=0$ the process described above can be carried out, but unfortunately we cannot make sure that the resulting manifold $\Sigma'^3$ is simply connected. However the Browder-Livesay invariant $\sigma(T, \Sigma^{4k+3})$ is clearly defined also when $\Sigma^{4k+3}$ is a homology sphere.

**Theorem.** *For every* $i \in \mathbb{Z}$ *there is a fixed point free involution* $(T, \Sigma^3)$ *of a homology sphere* $\Sigma^3$ *such that* $\sigma(T, \Sigma^3) = i$.

*Proof.* The only part of the proof given in II.4 that doesn't follow immetiately for the case $k=0$ is the construction of $A'$ in Lemma 1. The following "surgery" lemma fills this gap.

**Lemma.** *Let* $V$ *be the connected sum along the boundary of* $m$ *copies of* $S^1 \times D^2$ *and* $W = \partial V$. *Let* $H_1(W) = B_1 \oplus B_2$ *where* $B_1$ *and* $B_2$ *have rank* $m$ *and* $x \cdot y = 0$ *for all* $x, y \in B_1$. *Then* $W = \partial V'$, *where* $V'$ *is homeomorphic to* $V$ *and the map induced by the inclusion* $W \subset V'$ *sends* $B_1$ *to* $0$ *and* $B_2$ *isomorphically to* $H_1(V')$.

*Proof.* Let $\alpha_1, \ldots, \alpha_m, \beta_1, \ldots, \beta_m$ be the standard basis for $H_1(W)$ for which $\alpha_i \cdot \alpha_j = 0$, $\beta_i \cdot \beta_j = 0$ and $\alpha_i \cdot \beta_j = \delta_{ij}$, and $\alpha_1, \ldots, \alpha_m$ generate $K_1(W \to V)$. Let $\alpha'_1, \ldots, \alpha'_m$ be a basis for $B_1$ and $\beta'_1, \ldots, \beta'_m \in H_1(W)$ such that $\alpha'_i \cdot \beta'_j = \delta_{ij}$ and $\beta'_i \cdot \beta'_j = 0$. Then the automorphism of $H_1(W)$ given by $\alpha'_i \to \alpha_i$, $\beta'_i \to \beta_i$ preserves intersection numbers and can be lifted to an automorphism of $\pi_1(W)$ (see [38], pp. 177–178 and 355–356), and therefore by Nielsen's Theorem ([64], p. 266) can be realized by a homeomorphism $h: W \to W$. Then let $V'$ be the mapping cylinder of $h$, union $V$. $V'$ clearly has the required properties.

**Corollary.** *There exist infinitely many irreducible 3-manifolds with the same homology as* $P^3$.

*Proof.* Let $(T, \Sigma^3)$ be a fixed point free involution of a homology 3-sphere $\Sigma$. Then we have the exact sequence

$$1 \to \pi_1(\Sigma) \to \pi_1(\Sigma/T) \to \mathbb{Z}_2 \to 1$$

from which it follows that $H_1(\Sigma/T) \approx \mathbb{Z}_2$, and since $T$ must be orientation preserving $H_3(\Sigma/T) \approx \mathbb{Z}$ and $H_2(\Sigma/T) \approx 0$ by duality. (In fact, a map $\Sigma/T \to P^3$ can be constructed which induces isomorphisms in homology.) Express $\Sigma/T$ as a connected sum of irreducible manifolds, $\Sigma/T = M_1 \# M_2 \# \cdots \# M_n$. Clearly only one of them, say $M_1$, is a homology $P^3$, the rest being homology spheres. Then the double covering $\tilde{M}_1$ corresponding to the commutator subgroup of $\pi_1(M_1)$ is a homology sphere with an involution $T_1$, since $\Sigma = \tilde{M}_1 \# 2M_2 \# \cdots \# 2M_n$, and $\sigma(T_1, \tilde{M}_1) = \sigma(T, \Sigma)$ since we can assume that those homology spheres have been attached away from a characteristic submanifold. Since we can start with $\sigma(T, \Sigma)$ with $\sigma(T, \Sigma) = i$, any $i$, we have constructed infinitely many $(T_1, \tilde{M}_1)$ and infinitely many irreducible $M_1$.

*Remark.* The homology 3-spheres that admit fixed point free involutions with $\sigma \neq 0$ cannot be homeomorphic to the standard sphere, since

Livesay's theorem ([50]) asserts that any involution of $S^3$ is equivalent to the antipodal map.

## II.6 Manifolds with the Same Regular Homotopy Type as Line Bundles over Projective Spaces

Recall the following result, due to L. Siebenmann ([71]): Assume that an open manifold $W^{n+1}$ has the same regular homotopy type as $M^n \times R$, where $M^n$ is a closed manifold and $n \geqq 5$, and $\tilde{K}_0(\pi_1(M))=0$. Then $W^{n+1}=M_1^n \times R$, where $M_1^n$ is a closed manifold. We can ask whether the same result holds if we replace $M \times R$ by $E(\xi)$, where $\xi$ is a line bundle over $M$. The answer is negative:

**Theorem.** *Let $\eta$ be the canonical bundle over $P^{2k}$, $k>1$. Then there exists an open manifold $W^{2k+1}$ with the same regular homotopy type as $E(\eta)$, which is not the total space of any line bundle over a manifold.*

*Proof.* We just have to take $W=\Sigma^{2k+1}/T-$pt., where $\sigma(T, \Sigma^{2k+1}) \neq 0$, as given by Theorem II.1-A and B. Then $W$ is regularly homotopy equivalent to $P^{2k+1}-\text{pt.}=E(\eta)$. But $W$ is not the total space of any bundle over a manifold $M^{2k}$, for $M^{2k}$ would be homotopy equivalent to $P^{2k}$ and $\tilde{M}^{2k} \subset \Sigma^{2k+1}$ would be an invariant sphere for $(T, \Sigma^{2k+1})$.

From Theorem I.1.3 and Siebenmann's results, it can be shown that such examples cannot exist for line bundles over $P^{2k+1}$, $k>1$.

In a similar way Theorems I.1.3 and II.1.C and D provide examples and counterexamples to the attempted generalization of the relative version of Siebenmann's theorem to total spaces of line bundles.

## II.7 Invariant Codimension 2 Spheres

Theorem II.4.3 shows that for any $(T, \Sigma^{4k+3})$, a $2k$-connected characteristic submanifold $W$ can be thought of as a characteristic submanifold for another involution $(T', \Sigma'^{4k+3})$ with $\sigma(T', \Sigma')=0$, on which we can therefore perform equivariant surgery until we obtain a sphere $S^{4k+2}$. In other words we can write $(T|W, W)=(T'|S^{4k+2}, S^{4k+2})$ $\# M$ where $M$ is the connected sum of several copies of $S^{2k+1} \times S^{2k+1}$. Since $(T'|S^{4k+2}, S^{4k+2})$ admits an invariant $S^{4k+1}$ (Theorem I.1.3) which we can assume disjoint from the discs where $M$ and $M^*$ are attached, so does $W$. So we have proved

**Theorem.** *Any $(T, \Sigma^{4k+3})$ admits an embedded, invariant codimension 2 sphere $S^{4k+1}$.*

In the p.l. case we are asserting that there is an invariant $S^{4k+1}$ which is locally flat, and in the smooth case that it can be taken to be diffeomorphic to the standard sphere. In Chapter VI we will prove that such an invariant sphere must be diffeomorphic to the standard sphere, and that an invariant unknotted $S^{4k+1}$ can be found if, and only if, $\sigma(T, \Sigma^{4k+3})=0$.

Out of general position it is clear that any $(T, \Sigma^3)$ with $\Sigma^3$ a homology sphere, or in fact any 3-manifold, admits an invariant $S^1$, so no restrictions on $k$ are necessary in the theorem.

Another proof of this theorem can be given using Theorem I.2.2.

Chapter III

# Relations with Non-simply-connected Surgery Obstructions

## III.1 Normal Invariants

### III.1.1 Normal Maps, Cobordisms and Invariants

Let $X$ be a Poincaré space of formal dimension $n$ and $f\colon M^n \to X$ a map of degree 1 from a manifold $M$ into $X$ ([11, 82]). The objective of surgery is to transform $f$ and $M$ to obtain a manifold homotopy equivalent to $X$. To be able to apply the techniques of surgery one has to assume the existence of a stable bundle $\xi$ over $X$ and a bundle map $b\colon \nu_M \to \xi$ covering $f$, where $\nu_M$ is the stable normal bundle of $M$:

$$\begin{array}{ccc} \nu_M & \xrightarrow{b} & \xi \\ \downarrow & & \downarrow \\ M & \xrightarrow{f} & X \end{array}$$

Two such coverings, $(b_1, \xi_1)$ and $(b_2, \xi_2)$ will be considered equivalent if there is a bundle equivalence $c\colon \xi_1 \to \xi_2$ such that $c\,b_1 = b_2$.

**Definition.** A *normal map* $(f, b)\colon M^n \to X$ consists of a degree 1 map $f\colon M \to X$ together with an equivalence class of coverings $b\colon \nu_M \to \xi$.

Two normal maps $(f_i, b_i)\colon M_i \to X$, $i=0,1$, are called *cobordant* if there is a map $F\colon N^{n+1} \to X$, where $N$ is a cobordism between $M_0$ and $M_1$, such that $F|M_i = f_i$, and a bundle map $B\colon \nu_N \to \xi$ covering $F$, where $\nu_N$ is the stable normal bundle of $N$ (think of $N$ as embedded in $S^{n+K} \times I$ with $N \cap S^{n+K} \times \{i\} = M_i$) and $\xi$ is a bundle over $X$, such that $B|\nu_{M_i}\colon \nu_{M_i} \to \xi$ is equivalent to $b_i\colon \nu_{M_i} \to \xi_i$. We refer to $(F, B)$ as a *normal cobordism* between $(f_0, b_0)$ and $(f_1, b_1)$. The normal cobordism classes of normal maps $(f, b)\colon M \to X$ are called the *normal invariants* of $X$.

Similarly we can consider a Poincaré pair $(Y, X)$, normal maps $f\colon (N, \partial N) \to (Y, X)$, and normal cobordisms between them, and a relative version where one assumes that $f|\partial N$ is a (simple) homotopy equivalence, and normal cobordisms contain $s$-cobordisms between the boundaries.

All the above definitions can be interpreted both in the smooth (smooth manifolds and linear bundles) and in the p.l. (p.l. manifolds and p.l. bundles) categories, so we can talk of smooth normal invariants and p.l. normal invariants.

### III.1.2 $G/PL$, $G/0$

To represent the set of normal invariants of $X$, we need recall Spivak's theorem. A *Spivak normal fibration* of $X$ is a pair $(\nu, \alpha)$ where $\nu$ is a stable spherical fibration over $X$ (with fibre $S^{K-1}$, say) and $\alpha \in \pi_{n+K}(T(\nu))$ is such that $h(\alpha) = \phi[X]$, where $T(\nu)$ is the Thom space of $\nu$, $h\colon \pi_{n+K}(T(\nu)) \to H_{n+K}(T(\nu))$ is the Hurewicz homomorphism, $\phi\colon H_n(X) \to H_{n+K}(T(\nu))$ is the Thom isomorphism and $[X]$ the fundamental class of $X$ (local coefficients are assumed if $X$ is non-orientable).

**Theorem.** (Spivak [73], Wall [80]). *Every Poincaré space $X$ has a Spivak normal fibration $(\nu, \alpha)$. Given two Spivak normal fibrations $(\nu_0, \alpha_0)$, $(\nu_1, \alpha_1)$ of $X$ there is a unique fibre homotopy equivalence $h\colon \nu_0 \to \nu_1$ such that $T(h)_*(\alpha_0) = \alpha_1$.*

This theorem has a relative version which gives existence and uniqueness of Spivak normal fibrations for Poincaré pairs.

Let $G_n$ be the $h$-space of mappings $(S^{n-1}, *) \to (S^{n-1}, *)$ of degree $\pm 1$, $G = \lim_{n\to\infty} G_n$ under suspension and $B_G$ its classifying space ([74]). $G$ and $B_G$ can be considered as the structural $h$-space and classifying space for stable spherical fibrations. We also have the classifying spaces $B_{PL}$ and $B_0$ for stable p.l. bundles and stable linear bundles, respectively, and maps

$$B_{PL} \longrightarrow B_G$$

$$B_0 \longrightarrow B_G$$

whose fibres are $G/PL$ and $G/0$, respectively.

The Spivak normal fibration $(\nu, \alpha)$ of a Poincaré space $X$ has a classifying map

$$X \longrightarrow B_G.$$

A lifting of this map to $B_{PL}$ gives a p.l. bundle $\xi$ and a fibre homotopy equivalence $h\colon \xi \to \nu$. Pulling the homotopy class $\alpha$ to $T(\xi)$, and making it $t$-regular to $X$, gives us a normal map

$$\begin{array}{ccc} \nu_M & \xrightarrow{b} & \xi \\ \downarrow & & \downarrow \\ M & \xrightarrow{f} & X. \end{array}$$

Reciprocally, given a normal map as above, then $(\xi, T(b)_* \alpha_M)$ is a Spivak normal fibration for $X$, where $\alpha_M \in \pi_{n+K}(T(\nu_M))$ is obtained by collapsing the complement of a tubular neighborhood of $M^n \subset S^{n+K}$. Therefore we have a fibre homotopy equivalence $h: \xi \to \nu$, which gives a lifting of the classifying map of $\nu$ to $B_{PL}$ (cf. [16], #4). Similar results hold for liftings into $B_0$. Now the standard homotopy-cobordism argument shows that this correspondence induces a bijection between normal invariants and homotopy classes of liftings and we have

**Theorem** (Sullivan). *The set of p.l. (resp. smooth) normal invariants of the Poincaré space $X$ is non-empty if, and only if, the Spivak normal fibration of $X$ can be reduced to a p.l. (resp. linear) bundle. If the set of normal invariants is non-empty, then it is in one-to-one correspondence with $[X, G/PL]$ (resp. $[X, G/0]$).*

This one-to-one correspondence depends on the choice of a particular lifting which will correspond to the constant map. This theorem has a relative version: for a Poincaré pair $(Y, X)$ the set of normal invariants (if non-empty) is in one-to-one correspondence with $[Y, G/PL]$ or $[Y, G/0]$.

### III.1.3 Homotopy Triangulations and Smoothings

**Definition.** A *homotopy triangulation* (resp. *homotopy smoothing*) of the Poincaré space $X$ consists of a p.l. (resp. smooth) manifold $M^n$ and a simple homotopy equivalence $f: M^n \to X$. Two homotopy triangulations (resp. smoothings) $f_i: M_i \to X$, $i=0,1$, are *equivalent* if there is a p.l. homeomorphism (resp. diffeomorphism) $g: M_0 \to M_1$ such that $f_1 g \sim f_0$. The set of equivalence classes of homotopy triangulations (resp. smoothings) of $X$ will be denoted by $hT(X)$ (resp. $hS(X)$).

A homotopy triangulation or smoothing $f: M \to X$ determines a normal map, by taking $\xi = g^* \nu_M$, where $g$ is a homotopy inverse of $f$, and therefore a normal invariant $\alpha(f)$. $\alpha(f)$ will be called the normal invariant of $f$, or, sometimes, a normal invariant of $M$. By choosing a fixed normal invariant (usually through a fixed homotopy tirangulation or smoothing) we get maps

$$hT(X) \xrightarrow{\alpha} [X, G/PL],$$

$$hS(X) \xrightarrow{\alpha} [X, G/0].$$

The correspondence between normal invariants, homotopy triangulations and smoothings, and mappings into $G/PL$ and $G/0$ is natural under some circumstances. Let $g: N \to M$ be a map between manifolds, then to every normal invariant for $M$, we can associate a normal invariant for $N$ in the following cases:

a) $g$ is an embedding with normal bundle $\zeta$. Then if $f\colon M'\to M$ is a normal map covered by $b\colon \nu_{M'}\to\xi$, make $f$ $t$-regular at $N$, $f^{-1}(N)=N'$, we have a normal map $f_1=f|N'\colon N'\to N$, covered by $\nu_{N'}=\nu_{M'}|N'\oplus f_1^*\zeta\to\xi|N'\oplus\zeta$.

b) $N=\partial M$ and $g$ is the inclusion. Then to every normal map $f\colon M'\to M$ we can associate $f|\partial M'\to N$.

c) $N$ is the total space of a disc bundle $\zeta$ over $M$ and $g$ is the projection $E(\zeta)\to M$. Then to every normal map $f\colon M'\to M$ we can associate a normal map $E(f^*(\zeta))\to E(\zeta)$. The same thing can be done to the associated sphere bundle $\dot{E}(\zeta)$, by combining with case b).

d) $N=M\# W$ and $g$ collapses $W$ to a point. Then to every normal map $f\colon M'\to M$ we can associate the normal map $f\#1\colon M'\# W\to M\# W$.

In cases b), c) and d) the correspondence sends homotopy triangulations (resp. smoothings) to homotopy triangulations (resp. smoothings). Notice that in all cases we are assuming $M$ and $N$ to be manifolds, so we have a well defined bijection between normal invariants and mappings into $G/PL$ or $G/0$.

**Lemma.** *In the cases described above the induced map between normal invariants is given by the map* $g^*\colon [M, G/PL]\to[N, G/PL]$ *induced by* $g$. *In cases* b), c) *and* d) *this can be described by the commutative diagram*

$$\begin{array}{ccc} hT(M) & \xrightarrow{\alpha} & [M, G/PL] \\ {\scriptstyle g^*}\downarrow & & \downarrow{\scriptstyle g^*} \\ hT(N) & \xrightarrow{\alpha} & [N, G/PL] \end{array}$$

*Id.id. for* $hS(M)$, $hS(N)$ *and* $G/0$.

The proof follows from Spivak's theorem, and its relative version, by relating the various Thom spaces and fibre homotopy equivalences.

This lemma will be implicitly and explicitly used in the rest of this work.

## III.2 Non-simply-connected Surgery Obstructions

### III.2.1 The Groups $L_n(\pi,\omega)$

Let $(Y, X)$ be a Poincaré pair and $\pi=\pi_1(Y,*)$. There is a homomorphism $\omega\colon \pi\to\mathbb{Z}_2$ defined as follows: given $x\in\pi$, $\omega(x)=0$ if, and only if, the automorphism of the fibre over $*$ of the Spivak normal fibration of $(Y, X)$ obtained by translation along a loop representing $x$ is orientation preserving.

Now we can state the main results of [82].

First, to every pair $(\pi, \omega)$, $\pi$ a group and $\omega: \pi \to \mathbb{Z}_2$ a homomorphism, we can assign groups $L_n(\pi, \omega)$, periodic on $n$ of period 4. Given a normal map

$$f\colon (M^n, \partial M) \to (Y, X)$$

such that $f|\partial M: \partial M \to X$ is a simple homotopy equivalence, the *surgery obstruction* $\theta(f) \in L_n(\pi_1(Y), \omega)$ is defined. It depends only on the normal cobordism class of $f$ (rel. boundary), and is 0 if $f$ is a simple homotopy equivalence.

**Theorem A** (Wall [82]). *If $n \geqq 5$, $\theta(f)=0$ if, and only if, $f$ is cobordant (rel. boundary) to a simple homotopy equivalence.*

**Theorem B** (Wall [82]). *Given a closed manifold $M^n$, $n \geqq 5$, and $x \in L_{n+1}(\pi_1(M), \omega)$ there is a manifold $N^{n+1}$ with boundary $M \cup M_x$ and a normal map.*

$$\begin{array}{ccc} \nu_N & \longrightarrow & \nu_{M\times I} \\ \downarrow & & \downarrow \\ (N, M, M_x) & \xrightarrow{F_x} & (M \times I, M \times \{0\}, M \times \{1\}) \end{array}$$

*such that $F_x|M^n = 1$, $F_x|M_x^n = f_x\colon M_x \to M \times \{1\}$ is a simple homotopy equivalence, and $\theta(F_x) = x$. $f_x$ is a homotopy triangulation (smoothing) of $M$ which depends only on $x$.*

Now given a Poincaré complex $X$ and a given homotopy triangulation of $X$ these two theorems give us maps

$$[X, G/PL] \xrightarrow{\theta} L_n(\pi_1(X), \omega)$$
$$L_{n+1}(\pi_1(X), \omega) \to hT(X)$$

and we have an exact sequence

$$L_{n+1}(\pi_1(X), \omega) \to hT(X) \xrightarrow{\alpha} [X, G/PL] \xrightarrow{\theta} L_n(\pi_1(X), \omega).$$

And given a homotopy smoothing of $X$ we have an exact sequence

$$L_{n+1}(\pi_1(X), \omega) \to hS(X) \xrightarrow{\alpha} [X, G/0] \xrightarrow{\theta} L_n(\pi_1(X), \omega).$$

*Exactness* means that $L_{n+1}(\pi_1(X), \omega)$ acts on $hT(X)$ (or $hS(X)$) and $\alpha$ induces a bijection between the orbits of this action and the kernel of $\theta$. However, $\theta$ is not a homomorphism in general.

### III.2.2 The Obstruction Groups for $\pi = \mathbb{Z}_2$

When $\pi = \mathbb{Z}_2$ we have two possibilities for $\omega: \pi \to \mathbb{Z}_2$. The orientable case: $\omega = 0$, which we will denote by $(\mathbb{Z}_2, +)$, and the non-orientable case $\omega = 1$, which we will denote by $(\mathbb{Z}_2, -)$. The obstruction groups $L_n(\mathbb{Z}_2, \varepsilon)$, $\varepsilon = +$ or $-$, have been computed by Wall [81, 82] and are as follows:

| $n$ mod. 4 | 0 | 1 | 2 | 3 |
|---|---|---|---|---|
| $L_n(\mathbb{Z}_2, +)$ | $\mathbb{Z} \oplus \mathbb{Z}$ | 0 | $\mathbb{Z}_2$ | $\mathbb{Z}_2$ |
| $L_n(\mathbb{Z}_2, -)$ | $\mathbb{Z}_2$ | 0 | $\mathbb{Z}_2$ | 0 |

We need the explicit description of the surgery obstructions in $L_n(\mathbb{Z}_2, \varepsilon)$ when $n$ is even. Given a normal map $f: (N^{2k}, \partial N) \to (Y, X)$, where $(\pi_1(Y), \omega) = (\mathbb{Z}_2, \varepsilon)$, such that $f | \partial N$ is a simple homotopy equivalence, we can assume that $\pi_i(f) = 0$ for $i \leqq k$. Let $G = \pi_{k+1}(f)$. $G$ is a free $\Lambda$-module, where $\Lambda = \mathbb{Z}[\mathbb{Z}_2]$ is the group ring of $\mathbb{Z}_2$. Denoting by $t$ the generator of $\mathbb{Z}_2$, elements of $\Lambda$ will be formal linear combinations $x + yt$, $x, y \in \mathbb{Z}$. Given $\lambda = x + yt \in \Lambda$, let $\lambda^- = x + \varepsilon yt$. Then $\lambda \to \lambda^-$ is an involutory automorphism of $\Lambda$. Let $I$ be the subgroup of $\Lambda$ consisting of elements of the form $\lambda + (-1)^k \lambda^-$ and $V = \Lambda / I$. Then there are maps

$$\lambda: G \times G \to \Lambda$$

$$\mu: G \to V$$

satisfying:

(i) for $x \in G$ fixed, $y \to \lambda(x, y)$ is a $\Lambda$-homomorphism and the map $x \to \lambda(x, \ )$ is an isomorphism $G \to \mathrm{Hom}(G, \Lambda)$.

(ii) $\lambda(x, y) = (-1)^k \lambda(y, x)^-$.

(iii) $\lambda(x, x) = \mu(x) + (-1)^k \mu(x)^-$,

(iv) $\mu(x + y) = \mu(x) + \mu(y) + \lambda(x, y)$,

(v) $\mu(a x) = a^- \mu(x)\, a$, *for* $a \in \Lambda$.

The triple $(G, \lambda, \mu)$ is called a special Hermitian form. $\lambda$ and $\mu$ can be defined as follows: an element $x$ determines a prefered class of immersions $x: S^k \to N$ and a lifting $\tilde{x}: S^k \to \tilde{N}^{2k}$. Then

$$\lambda(x, y) = \tilde{x} \cdot \tilde{y} + (\tilde{x} \cdot T_* \tilde{y})\, t,$$

$$\mu(x) = \phi(\tilde{x}) + \psi(\tilde{x})\, t.$$

Here $T$ denotes the natural involution of $\tilde{N}^{2k}$, and $\psi$ is defined as in I.1.3, except that we do not assume that the representative sphere is embedded (this is possible if, and only if, $\phi(\tilde{x}) = 0$), but the definition makes sense for an immersed sphere.

Surgery can be performed if, and only if, there is a $\Lambda$-basis $e_1, \ldots, e_m, f_1, \ldots, f_m$ of $G$ such that $\lambda(e_i, e_j) = \lambda(f_i, f_j) = 0$, $\lambda(e_i, f_j) = \delta_{ij} + 0t$, $\mu(e_i) = \mu(f_i) = 0$ and the surgery obstruction can be identified with the algebraic obstruction to finding such a basis, which is given by index and Arf invariants.

Let $\{\alpha_i\}$ be a $\Lambda$-basis of $G$ and let

$$\lambda(\alpha_i, \alpha_j) = h_{ij} + k_{ij}t,$$

$$\mu(\alpha_i) = m_i + n_i t.$$

There are two cases.

Case I: $k = 2l$, $\varepsilon = +$. We consider the matrices $H_1 = (h_{ij} + k_{ij})$, $H_2 = (h_{ij} - k_{ij})$. These matrices are even, symmetric and unimodular, so we can define $\sigma_i = \frac{1}{8}$ Index $H_i$, $i = 1, 2$. The pair $(\sigma_1, \sigma_2)$ is the surgery obstruction. For geometric reasons it is more convenient to use the pair $(\sigma, \tilde{\sigma})$ where $\sigma = \sigma_1$ and $\tilde{\sigma} = \sigma_1 + \sigma_2$, and define

$$\theta(f) = (\sigma, \tilde{\sigma}) \in \mathbb{Z} \oplus \mathbb{Z} \approx L_{4l}(\mathbb{Z}_2, +).$$

$\sigma$ and $\tilde{\sigma}$ can be interpreted as the standard index obstructions of the maps $f: N \to Y$ and $\tilde{f}: \tilde{N} \to \tilde{Y}$. (Cf. V.8, A.)

We also define the map $\tau: L_{4l}(\mathbb{Z}_2, +) \to \mathbb{Z}$ by $\tau(\sigma, \tilde{\sigma}) = 2\sigma - \tilde{\sigma} (= \sigma_1 - \sigma_2)$ and $\tau(f) = \tau(\theta(f))$. Notice that if $f$ is a map between closed manifolds then $\tilde{\sigma} = 2\sigma$ and $\tau(f) = 0$, because in that case Index$(\tilde{Y}) = 2$ Index$(Y)$, Index$(\tilde{N}) = 2$ Index$(N)$, and $\sigma = \frac{1}{8}$ (Index$(N)$ − Index$(Y)$), $\tilde{\sigma} = \frac{1}{8}$ (Index$(\tilde{N})$ − Index$(\tilde{Y})$).

Case II: $k = 2l + 1$ or $\varepsilon = -$. In this case we consider the quadratic form $\mu_2$ over $G \otimes_\Lambda \mathbb{Z}_2$ given by

$$\mu_2(\alpha_i \otimes 1) = m_i + n_i \quad \text{mod. } 2.$$

Then $\theta(f) = c(\mu_2) \in \mathbb{Z}_2 \approx L_{2k}(\mathbb{Z}_2, \varepsilon)$.

Two special Hermitian forms of one of the types contained in Case II with the same rank and Arf invariant are equivalent.

## III.3 Relations with the Browder-Livesay Invariants

### III.3.1 $l_n: BL_n(\varepsilon) \to L_{n-1}(\mathbb{Z}_2, -\varepsilon)$

We consider again the situation of I.2:

$$f: (T', M'^n) \to (T, M)$$

an equivariant homotopy equivalence, $W$ a characteristic submanifold for $(T, M)$ and $W' = f^{-1}(W)$, where $M$ and $W$ are 1-connected. Then we have induced maps

$$\begin{array}{ccc} M'/T' & \xrightarrow{g} & M/T \\ \cup & & \cup \\ W'/T' & \xrightarrow{h} & W/T \end{array}$$

We can consider $g$ and $h$ as normal maps (see III.1.3) and we have two obstructions defined: $\sigma(f)$, the obstruction to making $g$ $h$-regular to $W/T$, i.e. the obstruction to changing $h$ into a homotopy equivalence by exchanging handles inside $W'/T'$ (ambient surgery), and $\theta(h)$, the abstract surgery obstruction.

We will assume $K_i(f)=0$ for $i<(n-1)/2$.

**Lemma 1.** *Assume $n=2m+1$ and let $\alpha_1, \ldots, \alpha_r$ be a basis for $K_m(f, W', W)$. Then $\alpha_1, \ldots, \alpha_r$ is a $\Lambda$ basis for $\pi_{m+1}(h)$ and with respect to it the special Hermitian form is*

$$\lambda(\alpha_i, \alpha_j)=0+(\alpha_i \cdot T'_* \alpha_j)\, t$$

$$\mu(\alpha_i)=0+\psi(\alpha_i)\, t.$$

*Proof.* We know already that $\alpha_1, \ldots, \alpha_r, T'_*\alpha_1, \ldots, T'_*\alpha_r$ is a basis for $K_m(W' \to W) \approx \pi_{m+1}(h)$ (we will identify these two groups) so the first result is trivial. For the second, since $\alpha_i \in K_m(W' \to A')$, where $W'=A' \cap T'A'$, we have $\alpha_i \cdot \alpha_j=0$ and $\phi(\alpha_i)=0$.

Thus it is natural to define a homomorphism

$$l_n\colon BL_n(\varepsilon) \to L_{n-1}(\mathbb{Z}_2, -\varepsilon)$$

given in terms of representative quadratic forms in the following way: given $\sigma \in BL_n(\varepsilon)$ let $(G, B, \psi)$ be a quadratic form with $\sigma(G, B, \psi)=\sigma$. Then $l_n(\sigma)$ is defined by the special Hermitian form $(G \otimes \Lambda, \lambda, \mu)$ given by

$$\lambda(x \otimes 1, y \otimes 1)=0+B(x, y)\, t$$

$$\mu(x \otimes 1)=0+\psi(x)\, t$$

and set $l_n=0$ when $n$ is even. Lemma 1 can be now restated as

**Lemma 1'.** $l_n(\sigma(f))=\theta(h)$.

**Theorem.** *For $n$ even, and for $n=4k+3$, $\varepsilon=+$, $l_n=0$. For $n=4k+3$, $\varepsilon=-$, and for $n=4k+1$, $\varepsilon=+$, $l_n$ is an isomorphism $\mathbb{Z}_2 \approx \mathbb{Z}_2$. For $n=4k+1$, $\varepsilon=-$, $l_n\colon \mathbb{Z} \to \mathbb{Z} \oplus \mathbb{Z}$ is given by $l_n(\sigma)=(\sigma, 0)$.*

*Proof.* All cases follow trivially from the definitions, except for the case $n=4k+3$, $\varepsilon=+$. This one follows from

**Lemma 2.** *Let $H=(h_{ij})$ be a symmetric, even, unimodular matrix over the integers and $\phi(e_i)=h_{ii}/2$ mod. 2, the associated quadratic form. Then the Arf invariant of $\phi$ is 0.*

*Proof.* $c(\phi)$ is clearly additive with respect to direct sums, and therefore depends only on the class of $H$ in the Grothendieck group of even, symmetric, non-singular bilinear forms. By the results of [58], [70], this group is free on two generators, the class of the bilinear form with matrix $\begin{pmatrix} 0 & 1 \\ 1 & 0 \end{pmatrix}$, for which clearly $c(\phi)=0$, and the class of the bilinear form with matrix $E_8$ (cf. II.4), and the fact that $c(\phi)=0$ in this case can be verified directly (for example, using the proof of Lemma 3, VI.3).

It is a curious fact that the result of this lemma is directly contained in the definition of the Arf invariant given in [35].

An immediate consequence of the theorem is

**Corollary.** *Let $f\colon (T_1, M_1^{4k+1}) \to (T_0, M_0^{4k+1})$ be an equivariant homotopy equivalence, where $M_0$ is a closed manifold, $\pi_1(M_0)=0$ and $T_0$ is orientation reversing. Then $\sigma(f)=0$.*

*Proof.* By the theorem $l_n(\sigma(f))=(\sigma(f), 0)$, but since $(\sigma(f), 0)$ is the obstruction for a map between closed manifolds, then $\tau(\sigma(f), 0)= 2\sigma(f)=0$, by III.2.2.

We have only considered the bounded case in the special situation of I.1.4, but it is clear that Lemma 1′ also holds there (as it would hold in the general bounded case). But the corollary doesn't (Theorem II.1-C, III.3.3).

### III.3.2 Ambient vs. Abstract Surgery Obstructions

III.3.1 illustrates some of the possibilities that may occur in comparing ambient and abstract surgery obstructions. Given a homotopy equivalence $f\colon M'^n \to M^n$, $N$ a codimension 1 submanifold and $N'=f^{-1}(N)$

$$\begin{array}{ccc} M' & \xrightarrow[\approx]{f} & M \\ \cup & & \cup \\ N' & \xrightarrow[g]{} & N \end{array}$$

we can ask whether we can change g into a homotopy equivalence by abstract or by ambient (i.e. inside $M'$) surgery, or not.

If $M$ and $N$ are 1-connected, then the fact that we have an embedded problem makes the abstract obstruction 0, and in fact the surgery can be done inside $M'$ ([12]).

In the cases considered by Theorems I.1.3, I.1.4 and I.2.2 several possibilities can occur:

If $n$ is even, then the abstract and ambient obstructions are 0.

If $n=4k+3$, $\varepsilon=+$ the abstract obstruction is always 0, but the ambient surgery obstruction can take any value.

If $n=4k+3$, $\varepsilon=-$ or $n=4k+1$, $\varepsilon=+$, then the two obstructions coincide and can take any value.

If $n=4k+1$, $\varepsilon=-$, then in the closed case the two obstructions are 0; in the bounded case the abstract surgery obstruction is restricted but can be different from 0, and the ambient obstruction vanishes if, and only if, the abstract obstruction does.

Other cases can be studied as in [15]. Cf. also VI.3 for the codimension 2 case.

### III.3.3 Proof of Lemma 2, II.4

Lemma 2, II.4 is in fact a restatement of Theorem III.3.1 for the case $n=4k+3$, $\varepsilon=+$.

Let $H$ and $H'$ be two unimodular even symmetric matrices of the same rank $m$. We consider the following two special Hermitian forms on the free $\Lambda$-module $G$ on $m$ generators $\alpha_1, \ldots, \alpha_m$

$$\lambda(\alpha_i, \alpha_j)=0+h_{ij}t \qquad \lambda'(\alpha_i, \alpha_j)=0+h'_{ij}t$$

$$\mu(\alpha_i)=0+(h_{ii}/2)\,t \qquad \mu'(\alpha_i)=0+(h'_{ii}/2)\,t.$$

Since both have Arf invariant 0 (Lemma 2, III.3.1), they are equivalent, and there is a basis $\alpha'_i$ of $G$ such that $\lambda(\alpha'_i, \alpha'_j)=\lambda'(\alpha_i, \alpha_j)$, $\mu(\alpha'_i)=\mu'(\alpha_i)$. If $\alpha'_i=(p_{ij}+q_{ij}t)\,\alpha_j$ (summation convention) we have

$$\begin{aligned}\lambda(\alpha'_i, \alpha'_j)&=\lambda((p_{ik}+q_{ik}t)\,\alpha_k, (p_{jl}+q_{jl}t)\,\alpha_l)\\&=(p_{ik}h_{kl}q_{jl}-q_{ik}h_{kl}p_{jl})+(p_{ik}h_{kl}p_{jl}-q_{ik}h_{kl}q_{jl})\,t\end{aligned}$$

by properties (i) and (ii) of the definition of special Hermitian form. Therefore

$$PHQ^t-QHP^t=0, \qquad \text{i.e., } PHQ^t \text{ is symmetric}$$

and

$$PHP^t-QHQ^t=H'.$$

$$\mu(\alpha'_i)=\mu((p_{ij}+q_{ij}t)\,\alpha_j)=\mu(p_{ij}\alpha_j)+\mu(q_{ij}t\,\alpha_j)+\lambda(p_{ij}\alpha_j, q_{ij}t\,\alpha_j)$$

by property (iv). By an inductive application of property (iv) we get

$$\mu(p_{ij}\alpha_j)=0+\tfrac{1}{2}p_{ij}h_{jk}p_{ik}t$$

$$\mu(q_{ij}t\,\alpha_j)=0-\tfrac{1}{2}q_{ij}h_{jk}q_{ik}t$$

and from properties (i) and (ii)

$$\lambda(p_{ij}\alpha_j, q_{ij}\,t\,\alpha_j)=p_{ij}h_{jk}q_{ik}+0t.$$

So

$$\begin{aligned}\mu(\alpha_i')&=p_{ij}h_{jk}q_{ik}+\tfrac{1}{2}(p_{ij}h_{jk}p_{ik}-q_{ij}h_{jk}q_{ik})t\\&=(PHQ^t)_{ii}+\tfrac{1}{2}(PHP^t-QHQ^t)_{ii}t.\end{aligned}$$

Therefore $(PHQ^t)_{ii}=0$ mod. 2, i.e. $PHQ^t$ is even and all the conditions in the definition of compatibility are fulfilled.

*Note.* In a similar way, Theorem III.1.1 is the basis of the proof of Theorem II.1-D given in II.3. We used there implicitly the fact that the two forms

$$\lambda(e,f)=1+0t \qquad \lambda'(e,f)=0+t$$

$$\mu(e)=\mu(f)=1+0t \qquad \mu'(e)=\mu'(f)=0+t$$

defined on the free $\Lambda$-module on two generators $e, f$, are equivalent, since both forms have Arf invariant 1.

### III.3.4 Proof of Theorem II.1-C

Given $(T, \Sigma^{4k+4})$, an invariant $S_0^{4k+3}$ and $i\in\mathbb{Z}$, we consider $Q_0=S_0/T$. By Wall's realization theorem III.1.B there is a normal cobordism $f:(W, Q_0, Q_1)\to(Q_0\times I, Q_0\times\{0\}, Q_0\times\{1\})$ such that $f|Q_1: Q_1\to Q_0\times\{1\}$ is a homotopy equivalence and $\theta(f)=(i, 0)$, and we can assume $\pi_i(f)=0$ for $i\leqq 2k$. So we have a $\Lambda$-basis $\{\alpha_i\}$ for $\pi_{2k+1}(f)$ for which

$$\lambda(\alpha_i, \alpha_j)=0+h_{ij}t$$

$$\mu(\alpha_i)=0+(h_{ii}/2)\,t$$

where $(h_{ij})$ is a non-singular, even, symmetric matrix of index $8i$. The double covering $\tilde{W}$ is a framed cobordism between $S_0$ and $S_1=\tilde{Q}_1$ and $\alpha_i, T_*\alpha_i$ is a basis of $H_{2k}(\tilde{W})=\pi_{2k+1}(f)$ and $\alpha_i\cdot\alpha_j=0$, $\alpha_i\cdot T_*\alpha_j=h_{ij}$. As in the proof of Theorem II.1-D (II.3), we can do surgery on the elements $\alpha_i$, obtaining a cobordism between $\tilde{W}$ and $S_0\times I$, which we can fill with $D_0^{4k+4}\times I$ to get a manifold $V$ with boundary $D_0\times\{0\}\cup\tilde{W}\cup D_0\times\{1\}$. $V\cup_{(T,\tilde{W})}V^*$ is, by the Mayer-Vietoris argument already used twice (II.3 and II.4), an $h$-cobordism and we can identify it equivariantly with $(T\times 1, \Sigma\times I)$. Then we can interpret $S_1$ as another invariant sphere for $(T, \Sigma)$, $\tilde{W}=V\cup V^*$ as a characteristic cobordism between $S_0$ and $S_1$, $K_{2k}=K_{2k}(\tilde{W}\to V)$ is the free group generated by $\{\alpha_i\}$ and $(\alpha_i\cdot T_*\alpha_j)=(h_{ij})$. So $\sigma(T, \Sigma, S_0, S_1)=i$, as required.

Chapter IV

# Combinatorial Classification of Involutions

## IV.1 Some Maps and Exact Sequences

We consider in this section the maps

$$\pi_n\colon S^n \to P^n$$
$$i_n\colon P^n \to P^{n+1}$$
$$j_n\colon P^n \to S^n$$

which are, respectively, the double covering, the natural inclusion and the map of degree 1 (pinching the complement of an open ball to a point). $i_n$ is the cofibre (mapping cone) of $\pi_n$, $j_{n+1}$ is the cofibre of $i_n$, and the composition $j_n \pi_n$ has degree 0 if $n$ is even, degree 2 if $n$ is odd.

**Lemma 1.** *If $X$ is an h-space, then for all $n \geqq 0$ the sequence*

$$\begin{array}{ccccccc}
\cdots \to [\Sigma P^{n+1}, X] & \xrightarrow{\Sigma i_n^*} & [\Sigma P^n, X] & \xrightarrow{\Sigma \pi_n^*} & \pi_{n+1}(X) & \xrightarrow{j_{n+1}^*} & [P^{n+1}, X] \to \\
\wr\wr & & \wr\wr & & \wr\wr & \xrightarrow{i_n^*} & [P^n, X] \xrightarrow{\pi_n^*} \pi_n(X) \\
\cdots \to [P^{n+1}, \Omega X] & \xrightarrow{i_n^*} & [P^n, \Omega X] & \xrightarrow{\pi_n^*} & \pi_n(\Omega X) & &
\end{array}$$

*is exact.*

*If $n$ is odd* $\operatorname{im} \pi_n^* \supset 2\pi_n(X)$ *and* $\ker j_{n+1}^* \supset 2\pi_{n+1}(X)$.

*If $n$ is even* $\operatorname{im} \pi_n^* \subset \pi_n(X)_2$ *and* $\ker j_{n+1}^* \subset \pi_{n+1}(X)_2$.

(Here $G_2$ denotes the subgroup of $G$ of elements of order 2.)

*Proof.* This exact sequence is the Puppe sequence ([69]) of the cofibration $S^n \to P^n \to P^{n+1}$. We use the fact that $X$ is an $h$-space to identify $[S^n, X] = \pi_n(X)$, and to forget about base points in general. ($X$ simple, would have been enough.)

If $n$ is odd, $\operatorname{im} \pi_n^* \supset \operatorname{im} \pi_n^* j_n^* = 2\pi_n(X)$. If $n$ is even, an element of $\operatorname{im} \pi_n^*$ is represented by a map $f\colon S^n \to X$ such that $fa = f$, where $a$ is the antipodal map of $S^n$, which has degree $-1$. Therefore $f \sim -f$ and $[f] \in \pi_n(X)$ must have order 2. These results applied to $\Omega X$ yield the corresponding results for $\ker j_{n+1}^*$, using the exactness of the sequence.

*Some* spaces satisfy the condition

$$(!)\quad \begin{cases} \text{If } n \text{ is odd im } \pi_n^* = 2\pi_n(X), \ker j_{n+1}^* = 2\pi_{n+1}(X). \\ \text{If } n \text{ is even im } \pi_n^* = 0, \ker j_{n+1}^* = 0. \end{cases}$$

which is stronger than the conclusion of Lemma 1. Examples of spaces satisfying (!) are Eilenberg-MacLane spaces, $B_0$, and $G/PL$, as we will see in the next section. If $X$ has only two non trivial homotopy groups in dimensions $n$ and $m$, it is easy to show that $X$ satisfies (!) if, and only if, the $k$-invariant of $X$ acts trivially on the cohomology of $P^{m+1}$ and $\Sigma P^m$, so it is easy to give examples of $h$-spaces (in fact, infinite loop spaces) that do not satisfy condition (!). (Cf. IV.2, proof of the lemma.)

We shall presently show by explicit examples, that $G/0$ does not satisfy condition (!).

Now consider the map $P^{2n+1} \to CP^n$ induced from the natural fibration $S^{2n+1} \to CP^n$ by identifying antipodal points in each fibre.

**Lemma 2.** *If* $\pi_i(X)=0$ *for all odd* $i \leqq 2n+1$, *then this map induces an epimorphism*

$$[CP^n, X] \to [P^{2n+1}, X] \to 0.$$

*Proof.* Since $P^{2n+1} \to CP^n$ is an $S^1$ bundle, its mapping cylinder is the associated disc bundle $E$, and the obstructions to extending a map $P^{2n+1} \to X$ to $E \sim CP^n$ lie in the groups $H^i(E, \dot{E}; \pi_{i-1}(X))$. The Thom isomorphism $H^i(CP^n) \xrightarrow{\approx} H^{i+2}(E, \dot{E})$ shows that these groups are always 0.

## IV.2 Computation of $[P^n, G/PL]$

We first recall the known homotopy properties of $G/PL$. First, its homotopy is periodic of period 4, given by

| $i$ mod. 4 | 0 | 1 | 2 | 3 |
|---|---|---|---|---|
| $\pi_i(G/PL)$ | $\mathbb{Z}$ | 0 | $\mathbb{Z}_2$ | 0 |

Representative normal maps $M^i \to S^i$ of the generators of these groups are given by the Milnor manifold $\overline{M}_0^{4k}$ with index 8, $k>1$ (Index 16, $k=1$) and the Kervaire manifold $\overline{K}^{4k+2}$, $k \geqq 0$ with Kervaire invariant 1.

Let $\mathbb{Z}_{(2)} = \mathbb{Z}[\frac{1}{3}, \frac{1}{5}, \ldots]$, the ring of rational numbers with odd denominator, and $\iota: \mathbb{Z} \to \mathbb{Z}_{(2)}$ the natural inclusion.

Let $Y$ be the two-stage Postnikov system with $\pi_2(Y)=\mathbb{Z}_2$, $\pi_4(Y)=\mathbb{Z}$ and $k$-invariant $\delta Sq^2$: $K(\mathbb{Z}_2, 2) \to K(\mathbb{Z}, 5)$. Let $Y_{(2)}$ be the two stage

Postnikov system with $\pi_2(Y_{(2)})=\mathbb{Z}_2$, $\pi_4(Y_{(2)})=\mathbb{Z}_{(2)}$ and $k$-invariant the composition $K(\mathbb{Z}_2, 2)\xrightarrow{\delta Sq^2} K(\mathbb{Z}, 5)\xrightarrow{\iota_*} K(\mathbb{Z}_{(2)}, 5)$.

Let

$$\Pi = Y \times \prod_{i>1} K(\mathbb{Z}_2, 4i-2)\times K(\mathbb{Z}, 4i)$$

$$\Pi_{(2)} = Y_{(2)} \times \prod_{i>1} K(\mathbb{Z}_2, 4i-2)\times K(\mathbb{Z}_{(2)}, 4i).$$

There is a map $\iota\colon \Pi \to \Pi_{(2)}$ such that the induced map $\iota_*\colon \pi_i(\Pi)\to \pi_i(\Pi_{(2)})$ is the identity $\mathbb{Z}_2 \approx \mathbb{Z}_2$ for $i=4k+2$ and $\iota\colon \mathbb{Z}\to\mathbb{Z}_{(2)}$ for $i=4k$.

**Theorem** (Sullivan [76]). *There is an h-map $G/PL\to\Pi_{(2)}$ (for some h-space structure of $\Pi_{(2)}$) such that the induced map $\pi_i(G/PL)\to\pi_i(\Pi_{(2)})$ is the identity $\mathbb{Z}_2\approx\mathbb{Z}_2$ for $i=4k+2$ and $\iota\colon\mathbb{Z}\to\mathbb{Z}_{(2)}$ for $i=4k$.*

This theorem asserts that $G/PL$ and $\Pi$ are equivalent mod. odd torsion groups. Therefore as viewed from $P^n$ they should look very similar, and $\Pi$ is much simpler than $G/PL$ to work with.

**Theorem.** *The maps $G/PL\to\Pi_{(2)}\leftarrow\Pi$ induce bijections*

$$[P^n, G/PL]\xrightarrow{\approx}[P^n, \Pi_{(2)}]\xleftarrow{\approx}[P^n, \Pi].$$

*Proof.* The second bijection follows from an easy computation (see the proof of next theorem), or by an argument similar to the one given below for the first one (since $\iota\colon\Pi\to\Pi_{(2)}$ can also be considered as an $h$-map).

We will prove that the first map is a bijection by induction on $n$, the result being clear for $n=1$. From Lemma 1, IV.1 we have the commutative diagram

$$\begin{array}{ccccccc}
\pi_{n+1}(G/PL) & \to & [P^{n+1}, G/PL] & \to & [P^n, G/PL] & \to & \pi_n(G/PL)\\
\downarrow & & \downarrow & & \downarrow & & \downarrow\\
\pi_{n+1}(\Pi_{(2)}) & \to & [P^{n+1}, \Pi_{(2)}] & \to & [P^n, \Pi_{(2)}] & \to & \pi_n(\Pi_{(2)})
\end{array}$$

The third vertical map is an isomorphism by induction hypothesis. We would like to show that the second one is also an isomorphism, and for that we need.

**Lemma.** *$\Pi_{(2)}$ satisfies condition (!) of* IV.1.

Assuming this lemma, for $n$ odd the above diagram reduces to

$$\begin{array}{cccccccc}
 & \mathbb{Z}_2 & \to & [P^{n+1}, G/PL] & \to & [P^n, G/PL] & \to & 0\\
 & \downarrow\wr & & \downarrow & & \downarrow\wr & & \\
0\to & \mathbb{Z}_2 & \to & [P^{n+1}, \Pi_{(2)}] & \to & [P^n, \Pi_{(2)}] & \to & 0
\end{array}$$

and for $n$ even to

$$\begin{array}{ccccccc} 0 \to [P^{n+1}, G/PL] & \longrightarrow & [P^n, G/PL] & \longrightarrow & \pi_n(G/PL) \\ \downarrow & & \downarrow \wr & & \downarrow \\ 0 \to [P^{n+1}, \Pi_{(2)}] & \longrightarrow & [P^n, \Pi_{(2)}] & \xrightarrow{0} & \pi_n(\Pi_{(2)}) \end{array}$$

and $\pi_n(G/PL) \to \pi_n(\Pi_{(2)})$ is in any case a monomorphism, and the conclusion $[P^{n+1}, G/PL] \approx [P^{n+1}, \Pi_{(2)}]$ follows from the five lemma.

*Proof of the Lemma.* All we have show is that each factor of $\Pi_{(2)}$ satisfies condition (!). This is trivial for a $K(\pi, n)$, so we just have to prove it for $Y_{(2)}$.

For $n=1$ the result is trivial, and for $n=2$ it follows from the fact that $Y_{(2)}$ is 1-connected and $j_2^*$ is onto, so $\operatorname{im} \pi_2^* = \operatorname{im} \pi_2^* j_2^* = 2\pi_2(Y_{(2)})$. For $n=4$ it is immediate from Lemma 1, IV.1 that $\operatorname{im} \pi_4^* \subset (\mathbb{Z}_{(2)})_2 = 0$. For $n=3$ it is necessary to show that $1 \notin \ker j_4^*$. For this we have an exact sequence

$$\begin{array}{ccccccc} [P^4, K(\mathbb{Z}_2, 1)] \xrightarrow{\delta Sq^2} & [P^4, K(\mathbb{Z}_{(2)}, 4)] & \longrightarrow & [P^4, Y_{(2)}] & \to [P^4, K(\mathbb{Z}_2, 2)] \to 0 \\ & \uparrow j_4^* & & \uparrow j_4^* & \\ & \pi_4(K(\mathbb{Z}_{(2)}, 4)) & \xrightarrow{\approx} & \pi_4(Y_{(2)}) & \end{array}$$

But in this case $\delta Sq^2 = 0$, the next map is a monomorphism, and since the result holds for $K(\mathbb{Z}_{(2)}, 4)$ it equally holds for $Y_{(2)}$.

**Corollary.** *$G/PL$ satisfies condition* (!) *of* IV.1.

**Theorem.** $[P^n, G/PL] \approx \mathbb{Z}_4 \oplus [(n/2) - 2]\, \mathbb{Z}_2$.

*Proof.*

$$[P^n, G/PL] \approx [P^n, \Pi] = [P^n, Y] \times \prod_{i>1} [P^n, K(\mathbb{Z}_2, 4i-2)] \times [P^n, K(\mathbb{Z}, 4i)]$$

and all the factors except the first are isomorphic to $\mathbb{Z}_2$. Now $[P^n, Y] = [P^4, Y]$, and from the exact sequence given above

$$0 \to \mathbb{Z}_2 \to [P^4, Y] \to \mathbb{Z}_2 \to 0$$

we see that $[P^4, Y]$ has order 4. To see it is actually $\mathbb{Z}_4$ we use the known fact that $[CP^2, G/PL] \approx \mathbb{Z}$ (in fact, this result is equivalent to the nontriviality of the first $k$-invariant of $G/PL$, [49]). From Lemma 2, IV.1

$$[CP^2, G/PL] \to [P^5, G/PL] \approx [P^4, G/PL] \approx [P^4, Y] \to 0$$

and therefore $[P^4, Y]$ is cyclic.

*Remarks.* We will denote by $\phi_i$ the component of the map $[P^n, G/PL] \to \mathbb{Z}_4 \oplus [(n/2)-2]\,\mathbb{Z}_2$ corresponding to dimension $i$:

$$\phi_4\colon\ [P^n, G/PL] \to [P^n, Y] \approx \mathbb{Z}_4$$

$$\phi_{4k}\colon\ [P^n, G/PL] \to [P^n, K(Z, 4k)] \approx \mathbb{Z}_2 \qquad k>1$$

$$\phi_{4k+2}\colon\ [P^n, G/PL] \to [P^n, K(Z_2, 4k+2)] \approx \mathbb{Z}_2$$

where $\phi_2$ is just $\phi_4$ reduced mod. 2. Therefore the values of $\phi_i$, $4 \leqq i \leqq n$ determine completely a normal invariant for $P^n$, and can be given arbitrarily.

These values can be detected by surgery obstructions. Let $x \in [P^n, G/PL]$ and $f\colon M^n \to P^n$ a normal map representing $x$ then $\phi_{4k+2}(x)$, $2 \leqq 4k+2 \leqq n$ can be computed as follows: make $f$ $t$-regular at $P^{4k+2} \subset P^n$, $N^{4k+2} = f^{-1}(P^{4k+2})$. Then $\phi_{4k+2}(x)$ is equal to the surgery obstruction (Arf-invariant) of the normal map $N^{4k+2} \to P^{4k+2}$. For $\phi_{4k}$, $4 \leqq 4k \leqq n$, the description is more elaborate: represent $[P^{4k}] \in H_{4k}(P^n, \mathbb{Z}_2)$ by a $\mathbb{Z}_2$-manifold $V^{4k} \to P^n$. Then $\phi_{4k}(x)$ is the surgery obstruction (Index/8 mod. 2 for $k>1$, mod. 4 for $k=1$) or "splitting invariant", of $f$ along $V^{4k}$. See [76] for the definition of these concepts.

Therefore the group structure of $[P^n, G/PL]$ given by the theorem comes from the addition of surgery obstructions along singular manifolds. This does not necessarily coincide with the addition coming from the natural $h$-space structure of $G/PL$. For example, these two structures are different for $[CP^n, G/PL]$ ([76]):

Another consequence of the description is that the top dimensional component $\phi_n$ of a normal map $f\colon M^n \to P^n$, $n$ even, can be changed by adding a Kervaire manifold if $n = 4k+2$ or a Milnor manifold if $n = 4k$.

A direct geometrical proof of the results of this section can be given, which amounts to the proof of Sullivan's theorem for this particular and simple situation.

## IV.3 The Classification Theorem

### IV.3.1 Homotopy Projective Spaces

**Lemma.** *Let $(T, \Sigma^n)$ be a fixed point free involution of the homotopy sphere $\Sigma^n$, and $Q^n = \Sigma^n/T$. Then there is a unique (up to homotopy) homotopy equivalence of degree 1, $Q^n \to P^n$. Therefore, the set of equivalence classes of p.l. (resp. smooth) involutions of homotopy n-spheres coincides with the set of homotopy triangulations (resp. homotopy smoothings) of $P^n$.*

*Proof.* Consider the set of maps $f\colon Q^n \to P^n$ that induce isomorphisms of $\pi_1$. If $n$ is odd, it follows from obstruction theory that the homotopy classes of such mappings are classified by their degree, which can be any odd integer. Therefore there is only one of degree 1. If $n$ is even we have to use obstruction theory with local coefficients, and in this case the (free) homotopy classes of such maps are classified by their absolute degree, which can be any odd positive integer, so again there is only one of degree 1. By considering now mappings $P^n \to Q^n$, the same argument shows that the homotopy equivalences $Q^n \to P^n$ are precisely those mappings of degree $\pm 1$ that induce isomorphisms of $\pi_1$, and the lemma is proved. (See [66], for details.)

### IV.3.2 Suspension

Let $(T, S^n)$ be a fixed point free p.l. involution of $S^n$. Applying construction II.2.1 to the pair $(D^{n+1}, S^n)$, we obtain a new involution $(T', S^{n+1})$ called the *suspension* of $(T, S^n)$. In terms of homotopy projective spaces, the suspension of $Q^n$ can be defined as the mapping cone of the double covering map $\tilde{Q}^n \to Q^n$, provided $\tilde{Q}^n = S^n$, which will always be the case for $n \geqq 5$. This defines a mapping

$$\Sigma\colon\ hT(P^n) \to hT(P^{n+1}) \qquad n \geqq 5.$$

Now we can consider the exact sequence of III.2.1 for the different values of $n$ mod. 4, together with the suspension map, the surgery obstruction groups as given by III.2.2, and the results of I.3, I.4, II.3 and IV.2. We can put all these together in a big diagram.

$$\begin{array}{ccccccc}
\mathbb{Z}_2 & \xrightarrow{0} & hT(P^{4k+1}) & \underset{\alpha}{\xrightarrow{\approx}} & [P^{4k+1}, G/PL] & \longrightarrow & 0 \\
 & & \wr\wr \big\downarrow \Sigma & & (\text{onto}) \big\uparrow i^*_{4k+1} & & \\
0 & \longrightarrow & hT(P^{4k+2}) & \underset{\alpha}{\longrightarrow} & [P^{4k+2}, G/PL] & \xrightarrow{\theta} & \mathbb{Z}_2 \longrightarrow 0 \\
 & & (\text{mono}) \big\downarrow \Sigma & & \wr\wr \big\uparrow i^*_{4k+2} & & \\
\mathbb{Z}\oplus\mathbb{Z} & \longrightarrow & hT(P^{4k+3}) & \underset{\alpha}{\longrightarrow} & [P^{4k+3}, G/PL] & \xrightarrow{\theta} & \mathbb{Z}_2 \\
 & & (\text{onto}) \big\downarrow \Sigma & & (\text{onto}) \big\uparrow i^*_{4k+3} & & \\
0 & \longrightarrow & hT(P^{4k+4}) & \underset{\alpha}{\longrightarrow} & [P^{4k+4}, G/PL] & \xrightarrow{\theta} & \mathbb{Z}_2 \\
 & & (\text{mono}) \big\downarrow \Sigma & & \wr\wr \big\uparrow i^*_{4k+4} & & \\
\mathbb{Z}_2 & \xrightarrow{0} & hT(P^{4k+5}) & \underset{\alpha}{\xrightarrow{\approx}} & [P^{4k+5}, G/PL] & \longrightarrow & 0
\end{array}$$

The squares are commutative, because $Q^n$ is a characteristic submanifold for its suspension (see Lemma III.1.3). The map $\mathbb{Z}_2 \to hT(P^{4k+1})$ is zero because a cobordism of the type described in Theorem B, III.2.1, can be obtained by adding a Kervaire manifold to $P^{4k+1} \times I$ along the boundary $P^{4k+1} \times \{1\}$ and therefore $P_x^{4k+1} = P^{4k+1} \# \Sigma^{4k+1} = P^{4k+1}$.

Also, $\theta$: $[P^{4k+2}, G/PL] \to \mathbb{Z}_2$ is onto because $\theta$ can be identified with $\phi_{4k+2}$. To determine $hT(P^n)$ for all $n \geqq 5$ we have to determine $\alpha$ and $\theta$ for $n=4k+3$ and $n=4k+4$. For this we need the results of the following section.

### IV.3.3 Two Useful Theorems

**Lemma.** *(Normal Cobordism Extension Property.) Let $M^n$ be a manifold and $N^q \subset M^n$ a submanifold with normal bundle $\zeta$. Let $f_0: M_0 \to M$ be a normal map, and suppose $f_0$ is t-regular at $N$, $N_0 = f_0^{-1}(N)$ and $g_0 = f|N_0$: $N_0 \to N$ the restricted normal map. Let $G$: $V^{q+1} \to N$ be a normal cobordism between $g_0$ and $g_1: N_1 \to N$. Then there is a normal cobordism $F$: $W^{n+1} \to M$ between $f_0$ and $f_1: M_1 \to M$ such that $F$ is t-regular at $N$, $F^{-1}(N) = V$ and $F|V = G$.*

*Proof.* By assumption we have bundle maps $b_0$: $\nu_{M_0} \to \xi$ covering $f_0$ and $B$: $\nu_V \to \zeta \oplus \xi|N$ covering $G$ such that $B|\nu_{N_0}$ coincides with the restriction of $b_0$, $\nu_{M_0}|N_0 \oplus g_0^* \zeta \to \xi|N \oplus \zeta$ (see III.1.3). So $\nu_V$ splits as $G^* \zeta \oplus G^*(\xi|N)$. Let $W = M_0 \times I \cup \bar{E}(G^* \zeta)$ attached along $\bar{E}(g_0^* \zeta) \times 1$

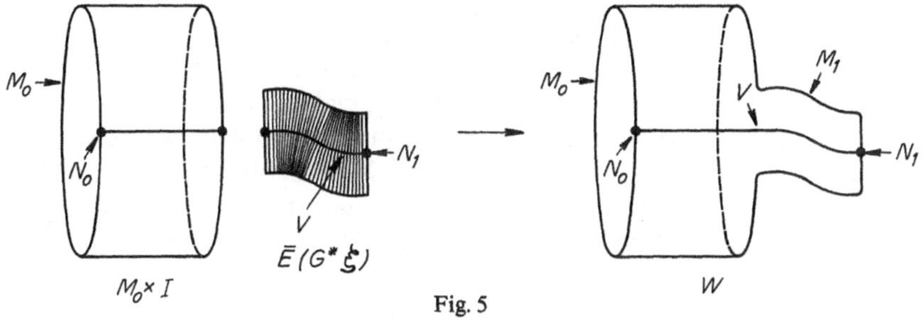


Fig. 5

where we have identified $\bar{E}(g^* \zeta)$ with a tubular neighborhood of $N_0$ in $M_0$. $W$ can be considered as a cobordism between $M_0 = M_0 \times \{0\}$ and $M_1 = (M_0 - \bar{E}(g^* \zeta)) \times \{1\} \cup \dot{E}(G^*(\zeta)) \cup \bar{E}(g_1^* \zeta)$ (with corners straightened in the smooth case). $F$: $W \to M$ is defined as $M_0 \times I \to M_0 \xrightarrow{f_0} M$ on $M_0 \times I$ and as $B|G^* \zeta$: $\bar{E}(G^* \zeta) \to \bar{E}(\zeta) \subset M$ on $\bar{E}(G^* \zeta)$. It is clear that $F$ is well defined, and since the normal bundle of $W$ is $\nu_M \times I \cup B^*(\xi|\bar{E}(\eta))$, also $b$ and $B$ can be combined to give a bundle map $\nu_W \to \xi$.

This lemma is used in [15], and, implicitly in [53], where the following applications were hinted at:

Let $M_1^{4k+4}$ be an unorientable manifold with fundamental group $\mathbb{Z}_2$, $M_2$ a characteristic submanifold for $M_1$ and $M_3$ a characteristic submanifold for $M_2$, i.e. $\tilde{M}_1 = A_1 \cup TA_1$, $A_1 \cap TA_1 = \tilde{M}_2 = A_2 \cup TA_2$, $\tilde{M}_3 = A_2 \cap TA_2$ and assume $\pi_1(M_2) \approx \pi_1(M_3) = \mathbb{Z}_2$. Then we have the inclusions $i_1\colon M_2 \to M_1$ and $i_2\colon M_3 \to M_2$ and the surgery obstruction maps $\theta_i\colon [M_i, G/PL] \to \mathbb{Z}_2$ (III.2.1 and III.2.2).

**Theorem 1.** *The following diagram is commutative:*

$$\begin{array}{ccc}
[M_1, G/PL] & & \\
\downarrow{\scriptstyle i_1^*} & \searrow{\scriptstyle \theta_1} & \\
[M_2, G/PL] & \xrightarrow{\theta_2} & \mathbb{Z}_2 \\
\downarrow{\scriptstyle i_2^*} & \nearrow{\scriptstyle \theta_3} & \\
[M_3, G/PL] & &
\end{array}$$

*Proof.* Assume first that $k>0$. To show that the upper triangle is commutative, it is enough to show that for every $x\in[M_1, G/PL]$, $\theta_1 x=0$ if, and only if, $\theta_2\, i_1^* x=0$. Let $f_1\colon N_1 \to M_1$ be a normal map representing $x$, $t$-regular at $M_2$, $N_2=f_1^{-1}(M_2)$, then $f_2=f_1|N_2\colon N_2 \to M_2$ represents $i_1^* x$. If $\theta_1(x)=0$, we can assume $f_1$ is a homotopy equivalence, and by I.2.4 we can make $f_1$ $h$-regular at $N_2$, so $\theta_2\, i_1^* x=0$. If $\theta_2\, i_1^* x=0$, then we can change $f_2$ into a homotopy equivalence through a normal cobordism, so by the above lemma, we can assume that $f_2$ is a homotopy equivalence, and we can further assume that $f_1|\bar{U}\colon(\bar{U}, \partial\bar{U}) \to (\bar{V}, \partial\bar{V})$ is a homotopy equivalence of pairs, where $U$, $V$ are tubular neighborhoods of $N_2$ in $N_1$ and $M_2$ in $M_1$, respectively. So we are left with the problem of making $f_1|N_1-U\colon N_1-U \to M_1-V$ a homotopy equivalence rel. boundary. Now $M_1-V=A_1$ is simply connected, so this surgery problem can be solved by adding a number of copies of the Milnor manifold $\bar{M}_0^{4k+4}$ to the interior of $N_1-U$. This has the effect of changing the normal invariant $x$ into $x'\in[M_1, G/PL]$ such that $\theta_1(x')=0$. But $\theta_1(x)=\theta_1(x')$, since the effect of this addition on the special Hermitian form of $f$ (III.2.2) consists in adding an even, symmetric, unimodular form, which has Arf invariant 0 (Lemma 2, III.3.1).

A similar argument applies for the lower triangle. Let $x\in[M_2, G/PL]$ be represented by $f_2\colon N_2 \to M_2$, $f_2^{-1}(M_3)=N_3$ and $f_3=f_2|N_3$ represents $i_2^* x$. If $\theta_2 x=0$, we can assume that $f_2$ is a homotopy equivalence and then $\theta_3(i_2^* x)=l_{4k+3}(\sigma(f_2))=0$ (III.2.1, Lemma 1′ and Theorem). If $\theta_3(i_2^* x)=0$ we can assume by the cobordism extension lemma that $f_3$ is a homotopy equivalence. Then, as before, we are left with the simply-connected surgery problem of transforming $f_2$ restricted to the complement of a tubular neighborhood of $N_3$ in $N_2$ into a homotopy equivalence, which can be solved because we are in an odd dimension. Therefore $\theta_2(x)=0$, and the lower triangle is commutative.

For $k=0$, by multiplying the manifolds $M_i$ by $CP^2$, we arrive at the situation already considered, but this does not change the obstructions (see [82], proof of Theorem 9.9).

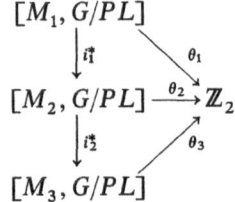

This theorem was proved by Wall [83] for $M_i^n = P^n$ by a completely different method, and the above argument for this same special case was given implicitly in [53]. The theorem also holds in the smooth case, since the obstruction maps factor through $[M_i, G/0] \to [M_i, G/PL]$.

Now we turn to the following situation: let $M^n$, $n \geqq 5$ be a manifold with $\pi_1(M^n) = \mathbb{Z}_2$, such that $M^n$ is orientable if $n$ is odd, unorientable if $n$ is even (just as in the case of $P^n$). If $n = 4k+3$ assume that there is a characteristic submanifold $N^{n-1}$ for $M^n(\tilde{M} = A \cup TA, \tilde{N} = A \cap TA)$ such that $\pi_1(N) \approx \mathbb{Z}_2$ and $(A, N)$ is 2-connected. (This condition can be removed: V.2.3 Corollary 1.)

**Theorem 2.** *Two homotopy triangulations of $M^n$ are equivalent if, and only if, they have the same normal invariant and the same Browder-Livesay invariant.*

*Proof.* Let $f_i$: $M_i \to M$ be two homotopy triangulations such that $\alpha(f_0) = \alpha(f_1)$ and $\sigma(f_0) = \sigma(f_1)$. From the first assumption it follows that there is a normal cobordism $F$: $W \to M$ between $f_0$ and $f_1$. If $n$ is even, we can make $W$ an $h$-cobordism since the surgery obstruction is 0 (III.2.2). If $n = 4k+1$ the same result holds, since the only obstruction is an Arf invariant, and we can change it by adding a Kervaire manifold to $W$ along the boundary. If $n = 4k+3$, compose $F$ with $f_0^{-1}$ to obtain a normal cobordism $f_0^{-1}F$ between 1: $M_0 \to M_0$ and $f_0^{-1}f_1$. Let $N_0$ be a characteristic submanifold of $M_0$ such that $\pi_1(N_0) = \mathbb{Z}_2$ and the pair $(A_0, \tilde{N}_0)$ is 2-connected, which we can obtain by making $f_0$ $t$-regular at $N$ and making $f_0|N_0$ sufficiently connected. Then, since $\sigma(f_0^{-1}f_1) = \sigma(f_0^{-1}) + \sigma(f_1) = 0$ (I.2.2, Lemma 2), we can assume $f_0^{-1}f_1$ $h$-regular at $N_0$, and set $N_1 = (f_0^{-1}f_1)^{-1}N_0$. Let $V = W$ be a (characteristic) cobordism joining $N_0$ and $N_1$. Since the surgery obstruction group is 0, we can make $V$ an $h$-cobordism, and by the normal cobordism extension lemma we can assume it is already an $h$-cobordism inside $W$. As in the proof of the previous theorem, we are left with the simply-connected surgery problem of obtaining a homotopy equivalence out of $f_0^{-1}F$ restricted to the complement of a tubular neighborhood of $V$ in $W$, and this can be solved by adding a number of copies of the Milnor manifold to $W$. So finally we obtain an $h$-cobordism, thus showing that in all cases the homotopy triangulations are equivalent.

The argument works also in the smooth case, giving

**Theorem 2'.** *Two homotopy smoothings of $M^n$ are equivalent, modulo the action of $\theta^n(\partial\pi)$, if, and only if, they have the same normal invariant and the same Browder-Livesay invariant.*

The argument can be used to show that if $M_1^n$ and $M_2^n$ are homotopy equivalent manifolds with the same normal invariant and $\pi_1(M_1^n) = \mathbb{Z}_2$, any dimension, any orientability, then $M_1^n \times S^2$ is p.l. homeomorphic to $M_2^n \times S^2$. Of course, this follows easily from non-simply-connected surgery (in fact for any fundamental group), but the point we want to make here is that the Browder-Livesay invariant $\sigma \in BL_{4k+3}(+)$ is a very weak (strong?) invariant: it can be changed arbitrarily within a normal invariant (V.2), and disappears when we cross a problem with $S^2$ or when we forget the ambient manifold (III.3.2). For $P^{4k+3}$, only its reduction mod. 2 can be detected after suspension (IV.4.2).

The proof of Theorems 2, 2′ extends easily to the cases $(4k, +)$ and $(4k+1, -)$, but examples can be constructed for the cases $(4k+2, +)$ and $(4k+3, -)$ where the proof definitely breaks down. However, we can use these same examples to prove Theorem 2 in these cases (but we do not know what happens in the smooth case). Since the obstructions lie in $\mathbb{Z}_2$, it is enough to exhibit a normal cobordism between $M^n$ and itself with non-zero surgery obstruction. For the case $(4k+2, +)$, let $N^{4k+1}$ be a characteristic submanifold with fundamental group $\mathbb{Z}_2$, and let $V^{4k+2}$ be a normal cobordism between $N$ and itself with Arf-invariant one, obtained, for example, by adding a Kervaire manifold to $N \times I$ along the boundary. The normal bundle $v$ of $N$ in $M$ pulls back to a bundle $\zeta$ over $V$, and $\dot{E}(\zeta) = \tilde{V}$ is a normal cobordism between $\tilde{N}$ and itself which has Arf-invariant 0, and therefore is normally cobordant, rel. boundary, to $\tilde{N} \times I$, by a normal cobordism $U$. Then $W = \bar{E}(\zeta) \cup U \cup (M - E(v)) \times I$ is easily seen to be a normal cobordism between $M$ and itself, and by a relative version of Theorem 1 (which can be proved in exactly the same way), has non-zero surgery obstruction. For the case $(4k+3, -)$, one simply applies the result just proved to a characteristic submanifold, and the result follows by an entirely analogous argument.

### IV.3.4 The Classification Theorem

**Theorem.** *For* $k>0$

$$hT(P^{4k+1}) \approx \mathbb{Z}_4 \oplus (2k-2)\,\mathbb{Z}_2$$

$$hT(P^{4k+2}) \approx \mathbb{Z}_4 \oplus (2k-2)\,\mathbb{Z}_2$$

$$hT(P^{4k+3}) \approx \mathbb{Z}_4 \oplus (2k-2)\,\mathbb{Z}_2 \oplus \mathbb{Z}$$

$$hT(P^{4k+4}) \approx \mathbb{Z}_4 \oplus (2k-1)\,\mathbb{Z}_2.$$

*Proof.* (See diagram IV.3.2.) We have shown before that $hT(P^{4k+1}) \approx [P^{4k+1}, G/PL]$. For $n = 4k+2$, $4k+4$ we have the exact sequence

$$0 \to hT(P^n) \xrightarrow{\alpha} [P^n, G/PL] \xrightarrow{\theta} \mathbb{Z}_2 \to 0$$

and $\theta$ can be identified in both cases with $\phi_{4k+2}$ (Theorem 1, IV.3.3). For $n=4k+3$ we have the following

**Lemma.** $hT(P^{4k+3})\approx hT(P^{4k+2})\times\mathbb{Z}$. ($k>0$.)

*Proof.* The bijection is given by $\beta(y)=(\alpha^{-1}i^*_{4k+2}\alpha(y),\sigma(y))$. Geometrically, the first component is obtained by taking a characteristic submanifold and doing surgery on it to obtain a homotopy $P^{4k+2}$. Since $[P^{4k+3},G/PL]\approx[P^{4k+2},G/PL]$, $\beta$ is a monomorphism by Theorem 2, IV.3.3. Now given $(x,\sigma)\in hT(P^{4k+2})\times\mathbb{Z}$, by Theorem II.4 there is $y\in hT(P^{4k+3})$ such that $\sigma(y)=\sigma$ and $y$ has a characteristic submanifold which is also a characteristic submanifold for $\Sigma x$. Therefore $i^*_{4k+2}\alpha(y)=\alpha(x)$, $\beta(y)=(x,\sigma)$ and $\beta$ is onto.

*Remarks.* Another proof of the Classification Theorem has been given by Wall [83], using instead of the Browder-Livesay invariant, another index invariant. We will see in IV.4 that these two invariants can be identified.

The numerical invariants in our classification are $\sigma$, and the compositions $\{\phi_i\alpha\}$ (see IV.2), which we will denote simply by $\phi_i$. Then our invariants are $\{\phi_i\}$, $4\leqq i\leqq 4k$ for $hT(P^{4k+1})$ and $hT(P^{4k+2})$; $\{\phi_i\}$, $4\leqq i\leqq 4k$, and $\sigma$ for $hT(P^{4k+3})$; and $\{\phi_i\}$, $4\leqq i\leqq 4k$ and $\phi_{4k+4}$ for $hT(P^{4k+4})$.

### IV.3.5 Proof of Theorem II.1-B (p.l. case)

From the classification theorem we have, for $k\geqq 0$, $hT(P^{4k+5})\approx\mathbb{Z}_4\oplus 2k\mathbb{Z}_2$, and $\sigma\colon hT(P^{4k+5})\to\mathbb{Z}_2$ can be identified with $\phi_{4k+2}$:

$$\sigma(x)=l_{4k+5}(\sigma(x))=\theta\, i^*_{4k+4}\alpha x=\phi_{4k+2}\alpha x=\phi_{4k+2}(x)$$

for $x\in hT(P^{4k+5})$, by III.3.1 (Lemma 1′ and Theorem), IV.3.3 (Theorem 1), and the above remarks.

Therefore exactly half of the p.l. involutions of $S^{4k+5}$, $k\geqq 0$, desuspend.

## IV.4 Another Relation with Non-simply-connected Surgery Obstructions

### IV.4.1 $\sigma=\pm\tau$

In Wall's version [83] of the classification theorem, an invariant $\tau$ is used, which for our purpose can be defined as follows: Let $Q^{4k+3}$ be a homotopy projective space, then from IV.3.4 there is a unique

$Q_0^{4k+3}$ such that $Q_0^{4k+3}$ has the same normal invariant as $Q^{4k+3}$, and $\sigma(Q_0^{4k+3})=0$. Let $F\colon W \to P^{4k+3}$ be a normal cobordism between $Q_0^{4k+3}$ and $Q^{4k+3}$, then define (III.2.2)

$$\tau(Q^{4k+3}) = \tau(F) = \tfrac{1}{8}(2 \operatorname{Index}(W) - \operatorname{Index}(\tilde{W})).$$

**Theorem.** *For $k>0$, $\tau(Q^{4k+3}) = \pm\sigma(Q^{4k+3})$. (Within a normal invariant the sign is constant.)*

*Remarks.* Hirzebruch and Jänich [33, 34] have shown that $\sigma=\tau$ in the smooth case. The proof uses the Atiyah-Bott-Singer fixed point theorem (see [3]), and it also shows that our definition of $\tau$ coincides with the one given there and in [83]. Wall has shown how to extend this proof to the p.l. case. Jänich and Ossa [36] have given an elementary proof of the main formula used in [33], [34]. Also results of Berstein and Livesay [6], and Orlik and Rourke [68] can be used to settle the sign.

The version of the result given here is all that is needed in the applications, and our proof has the advantage of being elementary (i.e. we use nothing we have not used so far) and it can be used to extend the result to involutions of more general manifolds (V.2). Actually, in [34] the result is proved for a very general situation. In V.8 we indicate the steps by which the result $\sigma=\tau$ can be proved in an even more general situation, mainly because we have a more general definition of $\sigma$ (I.2.2).

The idea behind the proof is the following: Wall's classification theorem and ours assert that within a normal invariant, homotopy projective spaces of dimension $4k+3$ are classified both by $\sigma$ and by $\tau$, so there is a bijection of $\mathbb{Z}$ to itself that sends $\sigma$ to $\tau$. We show that this bijection is a group homomorphism.

*Proof.* Given a homotopy projective space $Q_0^{4k+3}$ with $\sigma(Q_0^{4k+3})=0$, we construct a map

$$h\colon L_{4k+4}(\mathbb{Z}_2, +) \to BL_{4k+3}(+)$$

as follows: given $x \in L_{4k+4}(\mathbb{Z}_2, +)$, let $F_x\colon W_x \to Q_0 \times I$ be the cobordism between $1\colon Q_0 \to Q_0$ and $f_x\colon Q_x \xrightarrow{\simeq} Q_0$, such that $\theta(F_x)=x$, given by Wall's theorem B, III.2.1. Define then $h(x)=\sigma(f_x)=\sigma(Q_x)$.

We show now $h$ is a homomorphism. Let $x, y \in L_{4k+4}(\mathbb{Z}_2, +)$, and $F_x\colon W_x \to Q_0 \times I$, $F_y\colon W_y \to Q_0 \times I$, the corresponding normal cobordisms. Now we construct a homotopy equivalence $H\colon (W_y', Q_x, Q_{x,y}) \to (W_y, Q_0, Q_y)$ such that $H|Q_x = f_x$ (several alternative constructions of this are possible). Consider cobordisms (i) $F_{-x}\colon W_{-x} \to Q_0 \times I$ with $\theta(F_{-x}) = -x$, between $f_x$ and $1\colon Q_0 \to Q_0$, (ii) $F_{y,x}\colon W_{y,x} \to W_y \times I$ with $\theta(F_{y,x}) = x$, between $1\colon Q_y \to Q_y$ and $f_{y,x}\colon Q_{y,x} \to Q_y$.

We can combine these with $1: W_y \to W_y$

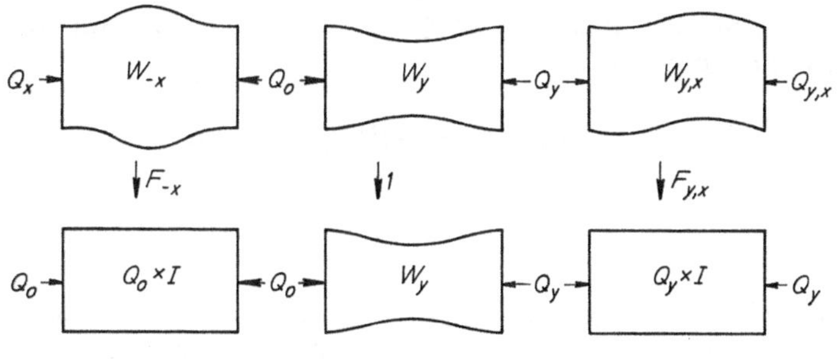


Fig. 6

to get a map $H'$: $W_{-x} \cup W_y \cup W_{y,x} \to W_y$. Since special Hermitian forms clearly add in this situation, $\theta(H') = \theta(F_{-x}) + \theta(1) + \theta(F_{y,x}) = 0$. Therefore we can make $H'$ into a homotopy equivalence $H$: $W'_y \to W_y$, keeping it fixed in the boundary.

Having constructed $H$: $W'_y \to W_y$, we can take $F_x \cup F_y H$: $W_x \cup W'_y \to Q_0 \times I$ as the normal map $F_{x+y}$: $W_{x+y} \to Q_0 \times I$ corresponding to $x+y$, because again we can add special Hermitian forms:

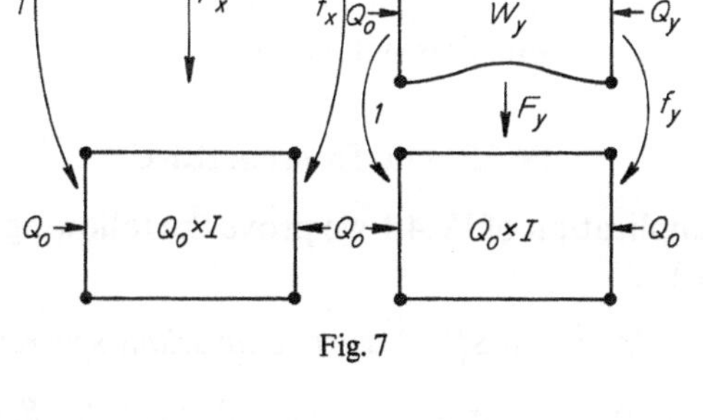

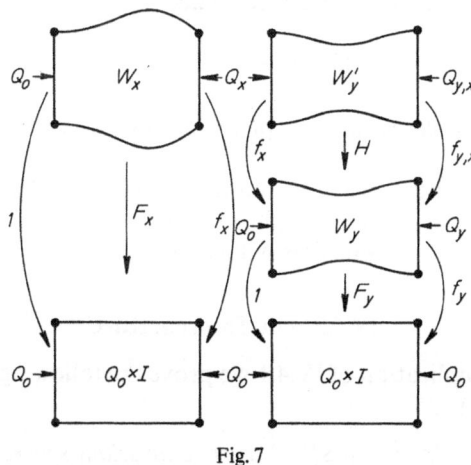

Fig. 7

Therefore

$$h(x+y) = \sigma(f_y \circ f_{y,x}) = \sigma(f_y) + \sigma(f_{y,x}) = \sigma(f_y) + \sigma(f_x) = h(x) + h(y),$$

by Lemmas 1 and 2, I.2.2, since $f_{y,x}$ and $f_x$ are cobordant through a homotopy equivalence.

To finish the proof of the theorem, we see that for elements of the form $x=(a, 2a)$, $h(x)=0$, since a representative for $W_x$ can be obtained by adding Milnor manifolds to $Q_0 \times I$ along $Q_0 \times \{1\}$, and $Q_x = Q_0$. Therefore $h$ factorizes as

$$\begin{array}{ccc} L_{4k+4}(\mathbb{Z}_2, +) & \xrightarrow{h} & BL_{4k+3}(+) = \mathbb{Z} \\ & {\scriptstyle \tau = 2\sigma - \tilde{\sigma}} \searrow \quad \nearrow {\scriptstyle h'} & \\ & \mathbb{Z} & \end{array}$$

By IV.3.4 (Lemma), $h$ is an epimorphism, and so is $h'$, and $h'$ must be an isomorphism, so $h'(x) = \pm x$. Therefore $\sigma(Q_x) = h(x) = h' \tau(x) = \pm \tau(x) = \pm \tau(Q_x)$. But every homotopy projective space with the same normal invariant as $Q_0$ must be equivalent to some $Q_x$.

*More remarks.* The definition of $h$ can be given in all the other cases, giving $h\colon L_{n+1}(\mathbb{Z}_2, \varepsilon) \to BL_n(\varepsilon)$, and the sequence

$$L_{n+1}(\mathbb{Z}_2, \varepsilon) \xrightarrow{h} BL_n(\varepsilon) \xrightarrow{l_n} L_{n-1}(\mathbb{Z}_2, -\varepsilon)$$

is exact in all cases. This can be seen as a reformulation of many of our results (II.1-A, III.3, V.2, etc.). Also Theorem 2, IV.3.3, is equivalent to the fact that the longer exact sequence

$$L_{n+1}(0) \to L_{n+1}(\mathbb{Z}_2, \varepsilon) \xrightarrow{h} BL_n(\varepsilon) \xrightarrow{l_n} L_{n-1}(\mathbb{Z}_2, -\varepsilon)$$

is exact in *some* cases, i.e. except for $n = 4k+2$, $\varepsilon = +$ and $n = 4k+3$, $\varepsilon = -$.

Notice that in the proof we did not actually use Theorem 2, IV.3.3: the fact that $h'$ is a monomorphism comes out directly from the fact that it is an epimorphism. This will come in handy in V.2.

### IV.4.2 On Theorem II.1-C

As a first application of IV.4.1 we prove the following strong version of Theorem II.1-C.

**Theorem.** *If $S_0^{4k+3}$ and $S_1^{4k+3}$ are two invariant spheres for $(T, \Sigma^{4k+4})$, then*

$$\sigma(T|S_1, S_1) - \sigma(T|S_0, S_0) = \pm 2\sigma(T, \Sigma, S_0, S_1).$$

*Therefore two desuspensions of $(T, \Sigma^{4k+4})$ are concordant if, and only if, they have the same Browder-Livesay invariant. Given a desuspension $(T|S_0, S_0)$, there is another desuspension $(T|S_1, S_1)$ with $\sigma(T|S_1, S_1) = \sigma$ if, and only if, $\sigma = \sigma(T|S_0, S_0)$ mod. 2.*

**Corollary.** *Two desuspensions of* $(T, \Sigma^{4k+4})$, $k>0$, *are concordant if, and only if, they are equivalent.*

*Proof.* By III.3 the abstract surgery obstruction of the characteristic cobordism $W$ between $S_0$ and $S_1$ is $(\sigma, 0)$, where $\sigma=\sigma(T, \Sigma, S_0, S_1)$. Therefore

$$\sigma(T|S_1, S_1)-\sigma(T|S_0, S_0)=\pm(\tau(S_1/T)-\tau(S_0/T))=\pm\tau(W)=\pm 2\sigma.$$

The rest of the theorem and the corollary now follow from Theorem II.1-C.

Berstein and Livesay [6] give a direct proof of this theorem that shows that the sign is always $+$. As remarked before, this also follows from Hirzebruch [33], [34].

We can finally express all desuspension results in the form of exact sequences, $k>0$:

$$0\to hT(P^{4k+1})\xrightarrow[\approx]{\Sigma} hT(P^{4k+2})\longrightarrow 0$$

$$0\to hT(P^{4k+2})\xrightarrow{\Sigma} hT(P^{4k+3})\xrightarrow{\sigma} \mathbb{Z}\to 0$$

$$0\to \mathbb{Z}\to hT(P^{4k+3})\xrightarrow{\Sigma} hT(P^{4k+4})\longrightarrow 0$$

$$0\to hT(P^{4k+4})\xrightarrow{\Sigma} hT(P^{4k+5})\xrightarrow{\sigma} \mathbb{Z}_2\to 0$$

where $\mathbb{Z}$ acts on $hT(P^{4k+3})$ by $(n, Q)\to Q'$, where $Q'$ has the same normal invariant as $Q$ and $\sigma(Q')=\sigma(Q)+2n$.

## IV.5 On the Topological Classification of Involutions

The recent advance in the theory of topological manifolds, due to Kirby and Siebenmann [44, 45, 46], makes possible a topological classification of fixed point free involutions on spheres, along the same lines as the p.l. classification obtained above. We describe here briefly this classification.

The main results of Kirby and Siebenmann that are needed are:

a) A complete description of the homotopy type of $G/\mathrm{Top}$; $G/\mathrm{Top}\sim\Omega^4 G/PL$. That is the homotopy groups of $G/\mathrm{Top}$ are exactly those of $G/PL$, but the exceptional $k$-invariant is killed: modulo odd torsion groups $G/\mathrm{Top}$ looks just like the product of Eilenberg-MacLane spaces. Thus

$$[P^n, G/\mathrm{Top}]\approx[n/2]\,\mathbb{Z}_2$$

and the mapping $[P^n, G/PL]\to[P^n, G/\mathrm{Top}]$ sends $\mathbb{Z}_4$ into the first $\mathbb{Z}_2$ summand, and all the $\mathbb{Z}_2$-summands isomorphically.

b) A theory of transverse regularity and surgery in the topological category, provided that the dimension of the manifolds and submanifolds involved is at least 5.

From b) it follows that $\sigma(T, S^{4k+3})$ can be defined for a topological involution $T$, and is a topological invariant. This last fact for p.l. involutions was first proved by A. Casson (unpublished) using different methods. Also our construction that gives the realization of all values of $\sigma$ can be carried out.

Since all the surgery results follow in the same way as in the p.l. case, the classification theorem for topological involutions is analogous to that given by Theorem IV.3.4, the only difference being that the $\mathbb{Z}_4$ summand should be changed to $\mathbb{Z}_2 \oplus \mathbb{Z}_2$. Thus in every dimension $\geqq 5$ the number of topological involutions is the same as that of p.l. involutions, but the natural map is neither injective nor surjective: The invariants $\phi_{2i}$, $i>2$, are topological invariants (a fact that follows directly from Sullivan's work [76], since they can be described as surgery obstructions along singular manifolds) and so is $\sigma$. But only the mod. 2 reduction of $\phi_4$ is a topological invariant, and so, for every p.l. involution, there is exactly another one which is combinatorially different but topologically equivalent to it. The new $\mathbb{Z}_2$ invariant for topological involutions is precisely the obstruction to the existence of a p.l. structure, so there are as many topological involutions which admit a p.l. structure as those which do not admit one. In short, "the Hauptvermutung and the triangulation conjecture both fail for free involutions on spheres of dimension $\geqq 5$" (Siebenmann). Explicit counterexamples to the Hauptvermutung for manifolds of dimension 5 can be given by Brieskorn involutions (V.4).

Chapter V

# Smooth Involutions

## V.1 General Remarks on Smooth Involutions of Spheres

One would like to classify smooth involutions of homotopy spheres in the way we did it in the previous chapter for p.l. involutions. But one cannot compute the set of normal invariants $[P^n, G/0]$, not even in terms of the (mostly unknown) homotopy groups of $G/0$. The most we can do is get an estimate as follows: let $a_i$ = order of $A_i = \pi_i(G/0)$, $a_i^{(2)}$ = order of $A_i \otimes \mathbb{Z}_2$. Then an inductive application of Lemma 1, IV.1 for $X = G/0$ gives

$$\operatorname{order}[P^{2k}, G/0] \leqq \prod_{i=1}^{2k} a_i^{(2)}$$

$$\operatorname{order}[P^{2k+1}, G/0] \leqq \prod_{i=1}^{2k} a_i^{(2)} \cdot a_{2k+1}$$

and equality would hold if, and only if, $G/0$ satisfied condition (!) of IV.1. But an example given in V.5 in dimension 9 shows that $G/0$ does not satisfy condition (!). So this upper bound can be divided by 2 for $n \geqq 10$.

From this we can get an estimate on $I_k$, the number of smooth involutions of homotopy $k$-spheres

$$I_{2k} \leqq \prod_{i=1}^{2k} a_i^{(2)} \qquad (k \neq 2)$$

$$\tilde{I}_{4k+3} \leqq \prod_{i=1}^{4k+2} a_i^{(2)} \cdot \theta_{4k+3} \qquad (k \neq 0)$$

$$I_{4k+1} \leqq \prod_{i=1}^{4k} a_i^{(2)} \cdot \theta_{4k+1}$$

where $\tilde{I}_{4k+3}$ denotes the number of involutions with $\sigma = 0$ ($I_{4k+3}$ is, of course, infinite: Theorem II.1-A) and $\theta_i$ = order of $\theta^i$. The estimates for $I_{2k}$ and $\tilde{I}_{4k+3}$ are always too large (see V.4) and can be divided by another factor of 2.

Another difference with the p.l. case is that we do not have a well defined suspension map $hS(P^n) \to hS(P^{n+1})$, for to define the suspension of $Q^n$ we have to attach a disc to $\tilde{Q}^n$, which we can do if, and only if, $\tilde{Q}^n \approx S^n$, and even in this case there are several ways of attaching it. To go around this difficulty we have to define a new set $hS_0(P^n)$:

**Definition.** We consider triples $(T, \Sigma^n, h)$, where $(T, \Sigma^n)$ is a fixed point free differentiable involution and $h: \Sigma^n \to S^n$ is a diffeomorphism. Two of these triples $(T_i, \Sigma_i^n, h_i)$ $i=0,1$, are *equivalent* if there is an equivariant diffeomorphism $f: \Sigma_0^n \to \Sigma_1^n$ such that $h_1 f h_0^{-1}: S^n \to S^n$ extends to a diffeomorphism of $D^{n+1}$. The set of equivalence classes of such triples will be denoted by $hS_0(P^n)$.

To relate $hS_0(P^n)$ and $hS(P^n)$ there is an exact sequence $(n \geqq 5)$

$$\theta^{n+1} \to hS_0(P^n) \to hS(P^n) \to \theta^n$$

where $\theta^{n+1}$ acts on $hS_0(P^n)$ by sending $(T, \Sigma^n, h)$ to $(T, \Sigma^n, gh)$, where $g: S^n \to S^n$ is a diffeomorphism representing the given element of $\theta^{n+1}$. The next map forgets the diffeomorphism $h$, and the last is the double covering.

By attaching $D^{n+1}$ to $\Sigma^n$ via $h$, and applying construction II.2.1 we obtain a suspension map

$$\Sigma: hS_0(P^n) \to hS(P^{n+1}).$$

This map preserves the action of $\theta^{n+1}$.

Then the desuspension results in the smooth case can be restated in the form of exact sequences $(k \geqq 1)$:

$$\mathbb{Z}_2 \to hS_0(P^{4k+1}) \xrightarrow{\Sigma} hS(P^{4k+2}) \longrightarrow 0$$

$$0 \to hS_0(P^{4k+2}) \xrightarrow{\Sigma} hS(P^{4k+3}) \xrightarrow{\sigma} \mathbb{Z} \to 0$$

$$0 \to \mathbb{Z} \to hS_0(P^{4k+3}) \xrightarrow{\Sigma} hS(P^{4k+4}) \longrightarrow 0$$

$$0 \to hS_0(P^{4k+4}) \xrightarrow{\Sigma} hS(P^{4k+5}) \xrightarrow{\sigma} \mathbb{Z}_2 \to 0$$

where the maps $\mathbb{Z}_2 \to hS_0(P^{4k+1})$ and $\mathbb{Z} \to hS_0(P^{4k+3})$ come from the action of $\mathbb{Z}_2$ and $\mathbb{Z}$ on those sets that can be best described by the proofs of Theorem II.1 C and D.

A few other results of Chapter IV can be pushed to the smooth case using these concepts.

## V.2 Normal Invariants and Browder-Livesay Invariants

### V.2.1 Involutions of Spheres

One main result relating the two invariants has already been given in Theorem 2′, IV.3.3, which shows that a smooth homotopy projective space of dimension $\geqq 5$ is determined, up to the action of $\theta^n(\partial\pi)$, by its normal invariant and its Browder-Livesay invariant.

For $n=4k+1$, $\sigma=\phi_{4k-2}$ (IV.3.5), so the Browder-Livesay invariant is determined by the normal invariant. In this section we will show that for $n=4k+3$, these two invariants are independent. This will give a stronger form of Theorem II.1-A.

**Theorem.** *Given a smooth homotopy projective space $Q^{4k+3}$, $k>0$, and $i\in\mathbb{Z}$, there is another one $Q_i^{4k+3}$ with the same normal invariant as $Q^{4k+3}$ such that $\sigma(Q_i^{4k+3})=i$.*

*Proof.* This result is a particular case of our next theorem, and is also a direct consequence of IV.4.1, but a direct proof helps us to understand better the construction of II.4. Consider the commutative diagram

$$\begin{array}{ccc} & & [P^{4k+2}, G/0] \\ & & \uparrow{\scriptstyle i^*} \\ hS(P^{4k+3}) & \xrightarrow{\alpha} & [P^{4k+3}, G/0] \\ \uparrow & & \uparrow{\scriptstyle j^*} \\ \theta^{4k+3} & \xrightarrow{\alpha} & \pi_{4k+3}(G/0) \to 0. \end{array}$$

Theorem II.4 asserts that there exists $Q_i'\in hS(P^{4k+3})$ with a common characteristic submanifold with $Q$, i.e. $i^*\alpha(Q_i')=i^*\alpha(Q)$, and with $\sigma(Q_i')=i$. Exactness at $[P^{4k+3}, G/0]$ (Lemma 1, IV.1) shows that there is $x\in\pi_{4k+3}(G/0)$ such that $j^*x+\alpha(Q_i')=\alpha(Q)$. If $\alpha(\Sigma)=x$, then $Q_i=Q_i'\#\Sigma$ has $\alpha(Q_i)=\alpha(Q)$ and $\sigma(Q_i)=\sigma(Q_i')=i$. (Cf. IV.3.4, Lemma.)

In the proof of IV.4 there was an element of indeterminacy in the construction, when a disc was attached at an $S^{4k+2}$ to get the manifold $A'$ (proof of Lemma 1). The above argument shows that we can attach this disc in such a way that we end up with a homotopy projective space with the same normal invariant as that of the one we started with. There is still an indeterminacy coming from $\theta^{4k+3}(\partial\pi)$. A related construction is given in [68].

### V.2.2 Involutions of Simply-connected Manifolds

**Theorem.** *Let $(T, M^{4k+3})$, $k>0$ be an orientation preserving involution of a simply-connected manifold $M$. Then, for every $i\in\mathbb{Z}$, there exists an equivariant homotopy equivalence $f\colon (T', M')\to(T, M)$, such that $\sigma(f)=i$ and $M'/T$ has the same normal invariant as $M/T$.*

*Proof.* Let $N=A\cap TA$ be a characteristic submanifold for $(T, M)$ which we can assume simply connected, and $H$ an even, symmetric, unimodular matrix of index $8i$ and rank $2r$. Let $(T', N')=(T, N)\# W$

(II.2.2), where $W$ is the connected sum of $r$ copies of $S^{2k+1} \times S^{2k+1}$, and we have an equivariant map $(T', N') \to (T, N)$ such that $K_{2k+1}(N' \to N)$ is a free group on generators $\alpha_i$, $T_* \alpha_i$, $i=1, \ldots, 2r$, and an equivariant normal cobordism $(T, X) \to (T, N)$ (i.e. $X/T \to N/T$ is a normal cobordism) between $(T', N') \to (T, N)$ and the identity. From the algebraic results of II.4 it follows that there exist elements $\alpha_i' \in K_{2k+1}(N' \to N)$ such that $\alpha_i' \cdot \alpha_j' = 0$, $\phi(\alpha_i') = 0$ and $(\alpha_i' \cdot T_* \alpha_j') = H$, so we can perform surgery on the elements $\alpha_i'$, obtaining a normal cobordism $h$: $U \to N$ between $N' \to N$ and a homotopy equivalence $N'' \to N$, $K_{2k+1}(U \to N)$ is free on $2r$ generators $a_i$, and $K_{2k+1}(N' \to N) \to K_{2k+1}(U \to N)$ is given by $\alpha_i' \to 0$, $T_* \alpha_i' \to a_i$. These two cobordisms can be combined into a cobordism $X \cup U \to N$ and we can perform surgery on it rel. $N \cup N''$, obtaining a normal cobordism $Y$ between $X \cup U \to N$ and an $h$-cobordism $Z \to N$.

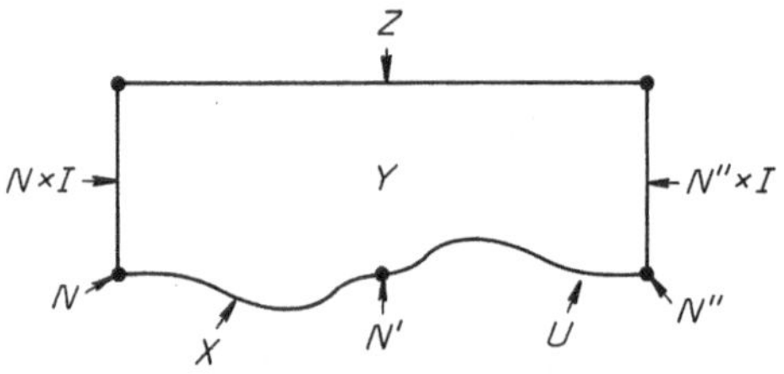


Fig. 8

This is so because $X \cup U$ is odd dimensional and $N$ is simply connected. (But even if $N$ were not simply connected this step can be carried out, because we can consider we are operating only in a small simply connected region of $N$; see remark below.)

We can now identify $N''$ with $N$, $Z \cup N'' \times I$ with $N \times I'$ and rearrange $Y$ and $G$ to get a map

$$G: (Y, X, N \times I') \to (N \times I, N \times \{0\}, N \times \{1\})$$

which restricted to $N \times I$ is the identity, restricted to $N \times I'$ is the projection into the first factor, and $G|X = h$

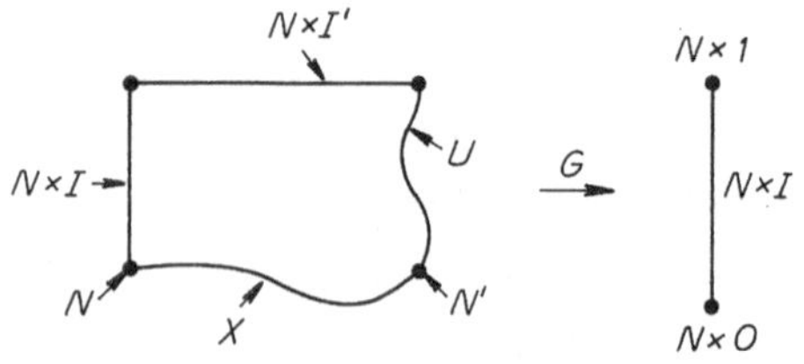


Fig. 9

Now we can construct maps

$$B = Y \cup A \times I' \to N \times I \cup A = A$$

which restricted to $Y$ is $G$ and restricted to $A \times I'$ is projection in the first factor, and finally

$$F\colon\ Q = B \cup_{(T,X)} B^* \to A \cup_{(T,N)} A^* = M$$

by construction II.2.1, since $G|X = h$ is equivariant.

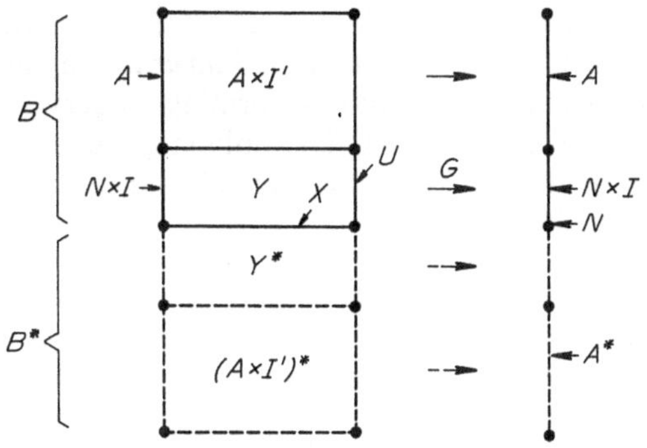


Fig. 10

$F$ is an equivariant cobordism between $1\colon M \to M$ and $f\colon M' = A' \cup_{(T',N')} A'^* \to M$, where $A' = U \cup A$.

Now we must show that $f$ is a homotopy equivalence. Clearly $K_i(A' \to A) = K_i(U \to N)$, and from the Mayer-Vietoris sequences of $(M'; A', A'^*)$ and $(M; A, A^*)$ we get an exact sequence

$$\cdots \to K_{i+1}(M' \to M) \to K_i(N' \to N) \to$$
$$K_i(A' \to A) \oplus K_i(A'^* \to A^*) \to K_i(M' \to M) \to \cdots$$

It follows as usual (cf. II.3, II.4) that the central map is always an isomorphism, so $K_i(M' \to M) = 0$ and $f$ is a homotopy equivalence.

Also $K_i(f) = K_i(N' \to N) \cap K_i(N' \to A')$ is generated by $\{\alpha_i'\}$ and $(\alpha_i' \cdot T_* \alpha_j') = H$ so $\sigma(f) = i$.

Finally $Q/T$ is a cobordism between $M'/T'$ and $M/T$ which we can describe as follows: let $\eta_N$ be the normal bundle of $N/T$ in $M/T$, and $\eta_X$ the normal bundle of $X/T$ in $Q/T$. Then $\dot{E}(\eta_N) = N$, $\dot{E}(\eta_X) = X$ and $M/T = E(\eta) \cup A$, $Q/T = E(\eta_X) \cup B$. Since $X/T \to N/T$ can be covered by a bundle map between the stable normal bundles, the same is true for $E(\eta_X) \to E(\eta_N)$, since $\eta_X$ is the pull-back of $\eta_N$. And in the construction of $B$ from $X$ we only used normal cobordisms, so indeed $Q/T \to M/T$ is a normal cobordism between $M'/T'$ and $M/T$.

*Remarks.* $(T', N')$ can also be considered as a characteristic submanifold for $(T, M)$, so the "common characteristic submanifold" result of II.4 holds here too. From the proof we can also add that $f$ is a diffeomorphism in the complements of appropriate neighborhoods of the characteristic submanifolds.

As remarked in the proof, the simple-connectivity assumption is not essential, and with more technical work the result can be proved for any connected, orientable manifold. An even more general form of this theorem appears as Lemma 12.10 in [82].

Also the proof can be applied to the case $T$ orientation reversing *in the p.l. case*, but *not within the normal invariant*, by adding Kervaire manifolds equivariantly to $N$ and performing surgery according to the algebraic blueprint given in II.3. The only step that cannot be carried out is the construction of the *equivariant* cobordism $h: X \to N$ (there is a cobordism). Therefore

**Theorem.** *Let $(T, M^{4k+3})$ be an orientation reversing p.l. involution of a simply-connected manifold $M$. Then there exists an equivariant homotopy equivalence $f: (T', M') \to (T, M)$ with $\sigma(f) = 1$.*

Again, simple-connectivity is not essential.

These two theorems and Corollary III.3.1, describe the values that $\sigma(f)$ can take in the general situation of Theorem II.2.2, except for the case $(4k+1, +)$, for involutions of closed manifolds. In the bounded case everything is easier, and all obstructions can be realized using Wall's Theorem III.2.1-B, as in III.3.4. The case $(4k+1, +)$ for closed manifolds is always the toughest one from our point of view, but an attack along the lines of [5] seems viable. With some additional hypotheses, the obstruction can be realized in this case by the construction given in V.5.

### V.2.3 $\sigma = \pm\tau$ again

Now we can define an invariant $\tau$ as in IV.4. Given $(T, M^{4k+3})$, where $M$ is simply-connected and $T$ is orientation preserving and $f: (T', M'^{4k+3}) \to (T, M^{4k+3})$ an equivariant homotopy equivalence. *Choose* an equivariant homotopy equivalence $f_0: (T_0, M_0) \to (T, M)$ which is normally cobordant to $f$ and such that $\sigma(f_0) = 0$, by V.2.2, (i.e. choose a fixed one with this property for each normal invariant of $M/T$, it will turn out at the end that the choice is irrelevant: Corollary 1), and let $F: (T, W) \to (T, M)$ be a normal equivariant cobordism between $f_0$ and $f$, i.e. $F/T$: $W/T \to M/T$ is a normal cobordism between $f_0/T_0$ and $f/T'$. Set $\tau(f) = \tau(\theta(F/T)) = \tau(F/T)$ (III.2.2).

**Theorem.** $\sigma(f) = \pm\tau(f)$.

With the understanding that the sign remains constant within a normal invariant.

The proof is identical with that in IV.4. As remarked there, the only fact needed is that $h'$ is onto, so we do not have to use Theorem 2, IV.3.3, which required additional conditions on $(A, \tilde{N})$.

**Corollary 1.** *Theorems* 2,2′, IV.3.3 *remain true without the connectivity condition on* $(A, \tilde{N})$.

*Proof.* The condition was needed only in the case $n=4k+3$. But assuming $\sigma(f_0)=\sigma(f_1)$, it follows that $\theta(F)$ is of the form $(a, 2a)$ and it can be killed by adding a Milnor manifold.

Finally, we can eliminate the connectivity conditions on Theorem I.2.2, by first proving the

**Lemma.** *Let* $f: (T_1, M_1^n) \to (T_0, M_0^n)$ *be an equivariant homotopy equivalence and* $N_0$ *a characteristic submanifold for* $(T_0, M_0^n)$. *Assume* $n$ *is odd* $\geqq 5$, *and* $M_0, N_0$ *are simply-connected. Then, if* $l_n(\sigma(f))=0$, *there is another equivariant homotopy equivalence with the same normal invariant as* $f$, *which can be made h-regular at* $N_0$.

*Proof.* The proof follows the lines of [11]. Let $N_1 = f^{-1}(N_0)$. Then $l_n\sigma(f)$ is the abstract surgery obstruction for $N_1/T_1 \to N_0/T_0$ (III.3.1), so there is a normal cobordism between it and a homotopy equivalence $N_1'/T_1' \to N_0/T_0$. By Lemma IV.3.3 this can be covered by a normal cobordism between $M_1/T_1 \to M_0/T_0$ and a normal map $f': M_1'/T_1' \to M_0/T_0$ such that $f'^{-1}(N_0/T_0) = N_1'/T_1'$. Now we can perform surgery on $f'$ outside $f'^{-1}(N_0/T_0)$ to make it a homotopy equivalence. (See proof of Theorem 1, IV.3.3.)

**Corollary 2.** *Theorem* I.2.2 *remains true without the connectivity conditions on* $(A_0, W_0)$.

*Proof.* The condition was needed only for $n$ odd. By the lemma, we can assume there is another equivariant homotopy equivalence $f'$: $(T_1', M_1') \to (T_0, M_0)$ with the same normal invariant as $f$ which can be made $h$-regular on $W_0$, and therefore $\sigma(f)=\sigma(f')=0$. If $T_0$ is orientation preserving, then Corollary 1 shows that $f$ and $f'$ differ only by the addition of a homotopy sphere, so $f$ itself can be made $h$-regular on $W_0$. If $T_0$ is orientation reversing and $n=4k+1$ the same result follows. If $n=4k+3$, $T_0$ orientation reversing, then the surgery obstruction of the normal cobordism between $f/T_1$ and $f'/T_1'$ might be non-zero in $L_{4k+4}(\mathbb{Z}_2, -)=\mathbb{Z}_2$. However, if this is the case, we can further extend it by a normal cobordism between $f'/T_1'$ and $f''/T_1''$ where $f''/T_1''$ is a homotopy equivalence $h$-regular at $W_0/T_0$, as in the proof of the lemma, but using a cobordism between the inverse images of $W_0/T_0$ with non-zero obstruction in $L_{4k+3}(\mathbb{Z}_2, +)$, using Theorem III.2.1-B. Now, an easily proved version of Theorem 1,IV.3.3 for the relative case shows that the cobordism between $f'/T_1'$ and $f''/T_1''$ has also non-zero obstruction, and finally the cobordism between $f/T_1$ and $f''/T_1''$ has obstruction 0, (the special Hermitian forms add), so $f$ is equivalent to $f''$, and $f$ is $h$-regular at $W_0$.

## V.3 Differentiable Structure of Spheres and Other Double Coverings

**Theorem.** *Assume $M^n$ is a manifold with fundamental group $\mathbb{Z}_2$, $n \geqq 5$, and $f_i$: $M_i^n \to M^n$, $i=0,1$ two homotopy smoothings of $M$ with the same normal invariant. If $M$ is orientable and $n=4k+3$, then $\tilde{M}_1 = \tilde{M}_0 \# m\,\Sigma_0$, where $\Sigma_0$ generates $\theta^{4k+3}(\partial\pi)$ and $m \equiv \sigma(f_1)-\sigma(f_0)$ mod. 2. In all other cases $\tilde{M}_1^n = \tilde{M}_0^n$.*

*Proof.* If $F$: $W \to M$ is a normal cobordism between $f_1$ and $f_0$, then $\tilde{F}: \tilde{W} \to \tilde{M}$ is a normal cobordism between $\tilde{M}_1$ and $\tilde{M}_0$, and since they are simply connected, $\tilde{M}_1 = \tilde{M}_0 \# m\,\Sigma_0$ where $m=\theta(\tilde{F})$. If $n=4k+3$ and $M$ is orientable, let $\theta(F)=(\sigma, \tilde{\sigma}) \in L_{4k+4}(\mathbb{Z}_2, +)$, and $\tilde{\sigma}=\theta(\tilde{F})$. But $\sigma(f_1)-\sigma(f_0) = \pm\tau(F) = \pm(2\sigma - \tilde{\sigma}) \equiv \tilde{\sigma}$ mod. 2. Looking at the other cases separately we see that $\theta(\tilde{F})=0$ always.

**Corollary.** *If a simply connected manifold $M^{4k+3}$, $k>0$, admits an orientation preserving fixed point free involution $T$, then every manifold of the form $M \# 2\Sigma_1 \# \Sigma_2$, where $\Sigma_1 \in \theta^{4k+3}$, $\Sigma_2 \in \theta^{4k+3}(\partial\pi)$, admits infinitely many different fixed point free involutions, all equivariantly homotopy equivalent to $(T, M)$.*

*Proof.* From the theorem and V.2.2 by varying $\sigma(f)$, $f$: $(T_1, M_1) \to (T, M)$, we can add arbitrary elements of $\theta^{4k+3}(\partial\pi)$, and by equivariant connected sum (II.2.2) we can add twice any element of $\theta^{4k+3}$. Since these groups are finite we obtain involutions of each $M \# 2\Sigma_1 \# \Sigma_2$ with infinitely many different values of $\sigma(f)$. To show all these are different, we proceed as follows: assume that $f_i/T_i$: $M_i/T_i \to M/T$ have the same normal invariant, and that there is a p.l. homeomorphism $h$: $M_1/T_1 \to M_0/T_0$. Let $F$: $W \to M/T$ be a normal cobordism between $f_1$ and $f_0$, and $W'$ obtained from $W$ by identifying $M_1/T_1$ and $M_0/T_0$ under $h$. Let $h' = f_0\, h\, f_1^{-1}$ and $M_{h'}$ the mapping torus of $h'$ (i.e., $M_{h'}$ is obtained from $M/T \times I$ by identifying $(x, 0)$ with $(h'(x), 1)$), and we get a degree 1 map $F'$: $W' \to M_{h'}$; since Index $(M_{h'})$ = Index $\tilde{M}_{h'}$ (= Index $M_{\tilde{h}'}$) = 0 ([13, 14]) we have $\sigma(\tilde{F}') = 2\sigma(F')$, and since $\theta(f) = (\sigma(F), \sigma(\tilde{F})) = (\sigma(F'), \sigma(\tilde{F}'))$ we get

$\tau(F)=0$, and therefore $\sigma(f_1)=\sigma(f_0)$. (Another proof of this result can be given by using the intrinsic Browder-Livesay invariant (I.2.2) or by cobordism theory as in [33, 83].)

In particular all $(4k+3)$-spheres of the form $2\Sigma_1+\Sigma_2$ admit fixed point free involutions, and in fact infinitely many.

Also we can say that all spheres of the form $2\Sigma_1$ admit involutions with all even values of $\sigma$, and those of the form $2\Sigma_1+\Sigma_0$ ($\Sigma_0$ = generator of $\theta^{4k+3}(\partial\pi)$) admit involutions with all odd values of $\sigma$, all of which can be obtained from the antipodal map by applying the construction II.4. A more direct proof of this has been found by Orlik and Rourke [68].

For 7-spheres, we can start with the curious involution of Hirsch and Milnor [31], which is an involution of $\Sigma_0$ with $\sigma=0$, and get all other possibilities:

**Corollary.** *Every homotopy 7-sphere admits an involution with any given value of $\sigma$.*

An independent proof has been given by Hirzebruch [33].

A better way to express these results is by defining the invariant $\kappa$:

Let $(T, \Sigma^n)$ be such that $\Sigma^n \in \theta^n(\partial\pi)$. If $\theta^n(\partial\pi)\neq 0$ we can write $\Sigma = m\,\Sigma_0$, where $\Sigma_0$ is a generator of $\theta^n(\partial\pi)$ and $m$ is well defined mod. 2. If $\theta^n(\partial\pi)=0$ set $m=0$. Then define

$$\kappa(T, \Sigma^n) = m + \sigma(T, \Sigma^n) \qquad \text{mod. } 2.$$

If $n=4k+1$ then both $m$ and $\sigma$ depend only on the normal invariant of $\Sigma^n/T$.

If $n=4k+3$, then $\kappa$ depends only on the normal invariant of $\Sigma/T$, and all values of $m$ and $\sigma$ compatible with $\kappa$ can be realized within a normal invariant.

Following Hirsch and Milnor [31], we can call *curious* an involution with $\kappa=1$. There are some obvious examples in the dimensions where the Kervaire invariant conjecture fails. But besides these, the example of Hirsch and Milnor, and related examples of Montgomery and Yang (see V.6), are the only curious involutions known to the author.

## V.4 Proof of Theorem II.1-B, Smooth Case

We consider the Brieskorn sphere $\Sigma_m^{4k+1} = \Sigma_{(m,2,\ldots,2)}^{4k+1}$, $m \geqq 3$ odd integer ([10, 32]). This is the submanifold of $\mathbb{C}^{2k+2}$ described by equations

$$z_0^m + z_1^2 + \cdots + z_{2k+1}^2 = 0$$

$$z_0\bar{z}_1 + z_1\bar{z}_1 + \cdots + z_{2k+1}\bar{z}_{2k+1} = 1.$$

$\Sigma_m^{4k+1}$ is a homotopy sphere and the involution $T: \Sigma_m^{4k+1} \to \Sigma_m^{4k+1}$ given by $T(z_0)=z_0$, $T(z_i)=-z_i$, $i>0$, is a fixed point free involution of $\Sigma_m^{4k+1}$.

**Theorem.** *If* $m \equiv \pm 1$ (mod. 8), *then* $\phi_{4i+2}(T, \Sigma_m^{4k+1})=0$ *for all* $i<k$. *If* $m \equiv \pm 3$ (mod. 8) *then* $\phi_{4i+2}(T, \Sigma_m^{4k+1})=1$ *for all* $i<k$.

*Proof.* Consider the following subsets of $\Sigma_m^{4k+1}$

$$W_1^{4k}=\Sigma_m^{4k+1} \cap \{\operatorname{Im} z_{2k+1}=0\}$$
$$W_2^{4k-1}=\Sigma_m^{4k+1} \cap \{z_{2k+1}=0\}$$
$$W_3^{4k-2}=\Sigma_m^{4k+1} \cap \{\operatorname{Im} z_{2k}=z_{2k+1}=0\}$$
$$\Sigma_m^{4k-3}=\Sigma_m^{4k+1} \cap \{z_{2k}=z_{2k+1}=0\}.$$

Then $\Sigma_m^{4k+1} \supset W_1 \supset W_2 \supset W_3 \supset \Sigma_m^{4k-3}$ is a chain of characteristic submanifolds. Now let $V^{4k-2}=W_3 \cap \{\operatorname{Re} z_{2k}>0\}$. Then $V^{4k-2} \subset D_+^{4k}= S^{4k} \cap \{\operatorname{Re} z_{2k}>0\}$ and $\partial V=\Sigma_m^{4k-3} \subset S^{4k-1}=\partial D_+^{4k}$. But according to Brieskorn [10], $(S^{4k-1}, \Sigma_m^{4k-3})$ is a knot with Arf invariant 0, if $m=\pm 1$ (mod. 8), or Arf invariant 1, if $m=\pm 3$ (mod. 8), which means that the Arf invariant of $V$ is 0 or 1, accordingly. $W_3/T \to P^{4k+2}$ represents the restriction of the normal invariant of $\Sigma_m^{4k+1}/T$ to $P^{4k+2}$, and its surgery obstruction gives the value of $\phi_{4k-2}(T, \Sigma_m^{4k+1})$. But this surgery obstruction can be identified with the Arf invariant of $V$. (Combinatorially $W_3/T=\Sigma(\Sigma_m^{4k-3}/T) \# \bar{V}$, where $\bar{V}=V \cup$ cone on $\partial V$.) This proves the theorem for $\phi_{4k-2}$, and continuing the argument for $\Sigma_m^{4k-3}$, etc., we get the theorem for $\phi_{4k-6}$, etc.

**Corollary 1.** *If* $m \equiv \pm 3$ *(mod. 8),* $\sigma(T, \Sigma_m^{4k+1})=1$.

Since, as in IV.3.5, $\sigma=\phi_{4k-2}$.

**Corollary 2.** *For all* $n \neq 4k+1$ *mod. 4 there exists* $x \in [P^n, G/0]$ *with* $\theta(x) \neq 0$.

This follows immediately from the theorem and Theorem 1, IV.3.3.

*Remarks.* In [17], Browder proves the theorem (and the corollaries) by showing that $(T, \Sigma_m^{4k+3})$ is equivariantly cobordant to $m$ copies of the antipodal map $(a, S^{4k+3})$, by a cobordism that preserves Arf invariants, and computing the Arf invariants of $m$ copies of $P^{4k+3}$ (see V.5). Giffen [25] gives a proof of this theorem that is similar to ours.

In the dimensions where the Kervaire invariant conjecture holds (i.e. for $4k+2 \neq 2^q-2$, at least, [16]) then the conclusion of the theorem also follows from the following consequence of V.3: Let $x \in [P^n, G/0]$

and $x'$ the restriction of $x$ to $P^{4k+1} \subset P^n$. Since the surgery obstruction group is 0, we can represent $x'$ by a homotopy projective space $Q^{4k+1}$. Then $\phi_{4k+2}(x)$ is 0 if, and only if, $\tilde{Q}^{4k+1}$ is diffeomorphic to the standard sphere.

Notice that $\kappa = 0$ for these examples, except for the obvious cases.

These involutions have been studied in great detail by Atiyah [3], Browder [17] and Giffen [25], [26], [27]. See also [35].

## V.5 Involutions and the Generalized Kervaire Invariant

### V.5.1 The Generalized Kervaire Invariant

We summarize here the main ideas and results of [17]. Let $\gamma^n$ be a bundle over a complex $B$. A $B$-orientation of a closed $m$-manifold $M$ is a bundle map $\nu_M \to \gamma$, where $\nu_M^n$ is the stable normal bundle of $M\,(n \gg m)$. $M$, together with a $B$-orientation is called a $B$-manifold, and one can clearly define the $B$-cobordism group of $m$-manifolds. Embed $B$ into $S^Q$ for some large $Q$, and let $B_1$ be a regular neighborhood of $B$ in $S^Q$, $\gamma_1^{-1}$ the inverse of the pull-back of $\gamma$ to $B_1$, $W = \bar{E}(\gamma_1^{-1})$. Let $p\colon W \to B_1$ be the projection.

Now let $M$ be a $B$-manifold, with a $B$-orientation covering a map $M \to B$. Let $c\colon M \to W$ be its composition with $B \subset W$. Then $c$ is covered by the bundle map $\tau(M) + \varepsilon^l + \varepsilon^Q \to p^*(\gamma_1^{-1}) + \varepsilon^Q = \tau(W)$, which determines a regular homotopy class of immersions $M^m \times D^k \to W$, and, up to an isotopy, an unique embedding $M^m \times D^k \subset W$. This defines, by the Pontryagin-Thom construction, a map

$$g\colon\ W/\partial W \to \Sigma^k M_+$$

(where $M_+ = M \cup \{\text{point}\}$) called the *spectral orientation map* of $M$ and is well defined up to homotopy. Composing $g$ with $\Sigma^k(c)_+\colon\ \Sigma^k M_+ \to \Sigma^k W_+$ we obtain a map

$$h = (\Sigma^k c)\, g\colon\ W/\partial W \to \Sigma^k W_+$$

called the *spectral cobordism invariant* of $M$. As its name indicates, the homotopy class of $h$ depends only on the $B$-cobordism class of $M$.

If $m = 2q$, let

$$\eta^*\colon\ H^q(M; \mathbb{Z}_2) \to H^{q+k}(W/\partial W, \mathbb{Z}_2)$$

be the composition

$$H^q(M; \mathbb{Z}_2) \xrightarrow{\Sigma^k} H^{q+k}(\Sigma^k M_+; \mathbb{Z}_2) \xrightarrow{g^*} H^{q+k}(W/\partial W; \mathbb{Z}_2)$$

and let $A = \ker \eta^*$.

Assume from now on that the $Wu$ class $v_{q+1}(\gamma)=0$. Then the quadratic form

$$\psi\colon A\to\mathbb{Z}_2$$

can be defined as follows: if $x\in A$ is represented by a map $\alpha\colon M_+\to K(\mathbb{Z}_2;q)$, let $\beta$ denote the composition

$$W/\partial W\xrightarrow{g}\Sigma^k M_+\xrightarrow{\Sigma^k\alpha}\Sigma^k K(\mathbb{Z}_2,q)$$

then the functional square $Sq_\beta^{q+1}(\Sigma^k(\iota))\in H^{2q+k}(W/\partial W;\mathbb{Z}_2)$ is well defined and define

$$\psi(x)=Sq_\beta^{q+1}(\Sigma^k(\iota))[W].$$

$\psi$ is quadratic: $\psi(x+y)=\psi(x)+\psi(y)+(x\cup y)[M]$. Therefore if the form $(x\cup y)[M]$ restricted to $A$ is non-singular, the Arf invariant of $\psi$ is defined. It can actually be defined in a more general situation: let $R=\{x\in A\,|\,(x\cup y)[M]=0$ for all $y\in A\}$; $\psi$ is called regular if $\psi|R=0$ and then $\psi$ induces $\psi'\colon A/R\to\mathbb{Z}_2$ whose associated bilinear form is non-singular, and one can define $c(\psi)=c(\psi')$. If $\psi$ is regular then we can define the (generalized) Kervaire invariant of $M$ by

$$k(M)=c(\psi).$$

We will be interested in the case of a normal map $f\colon M'^{2q}\to M^{2q}$ covered by $b\colon \nu_{M'}\to\xi$, where we can take $B=M$, $\gamma=\xi$ and consider $b$ as an $M$-orientation of $M'$. Since $v_{q+1}(\xi)=v_{q+1}(\nu_M)=0$, $\psi$ is well defined. In fact we have

**Theorem** (Browder [17]). *With $(f,b)$ defining the $M$-orientation of $M'$, $k(M')$ is defined. If $q$ is odd and $M$ is 1-connected, or if $\pi_1(M)=\mathbb{Z}_2$ and $M$ is non-orientable, then $k(M')=\theta(f)$.*

### V.5.2 Odd Multiples of an Involution

Let $M$ be a manifold with $\pi_1(M)=\mathbb{Z}_2$, $p\colon\tilde M\to M$ the double covering and $f\colon M'\to M$ a normal map covered by $b\colon\nu_{M'}\to\nu_M$ and $n=2s+1$ and odd integer. Form the manifold

$$M'(n)=n\,M'-s\,\tilde M$$

where the right hand side denotes the disjoint union of $n$ copies of $M'$ and $s$ copies of $-\tilde M$. Then there is a normal map

$$nf\colon\ M'(n)\to M$$

defined to be $f$ on each copy of $M'$ and $p$ on each copy of $-M$, and covered by $b$ on each $\nu_{M'}$ and by a fixed map $d$ covering $p$ on each $\nu_{\tilde M}$. (Clearly $nf$ has degree 1.)

This definition can be extended to the case of $n=2s+1$ normal maps

$$f_i\colon M_i' \to M \qquad i=1,\dots,n$$

covered by $b_i\colon \nu_{M_i'} \to \nu_M$, to give a normal map

$$a(f_1,\dots,f_n)\colon a(M_i',\dots,M_n') \to M$$

where $a(M_1',\dots,M_n') = \bigcup M_i' \cup (-s\tilde{M})$.

The Kervaire invariant of $M'(n)$ has been computed by Browder in some cases, one of which is the following:

**Theorem** (Browder [17]). *Assume that* $\dim M = 2q$ *and*

a) $0 = h_N^*\colon H^{q+k}(\Sigma^k M_+;\mathbb{Z}_2) \to H^{q+k}(\Sigma^k M_+;\mathbb{Z}_2)$,

*where $h_N$ is the spectral cobordism invariant of $N=\tilde{M}$ with respect to the $M$-orientation given by $d$.*

b) $k(N)$ is defined and $k(N)=0$

c) $v_q(M)^2 \neq 0$,

*then $k(M'(n))$ is given by*

$$k(M'(n)) = \begin{cases} k(M') & \text{for } n \equiv \pm 1 \ (8) \\ k(M')+1 & \text{for } n \equiv \pm 3 \ (8). \end{cases}$$

This result clearly applies when $M$ is homotopy projective space. As a first application we prove:

**Theorem.** *The set of homotopy $n$-spheres ($n \geqq 5$) which admit involutions is the subgroup of $\theta^n$ consisting of those spheres whose normal invariant lies in*

$$\pi_n^*[P^n, G/0] \subset \pi_n(G/0).$$

*Proof.* We will first show that if $x \in \pi_n^*[P^n, G/0]$, $x = \pi_n^*(x')$, then some sphere with normal invariant $x$ admits an involution. For $n=4k+1$ this is trivial, since $x'$ can be represented by a homotopy projective space. In the other case represent $x'$ by a normal map

$$f\colon M' \to P^n.$$

If $\theta(f)=0$ then $x'$ can be represented by a homotopy projective space, and the assertion follows. If $\theta(M')=1$ form

$$3f\colon M'(3) \to P^n$$

and $\theta(3f)=0$ by the previous theorem, since $\theta(3f)=k(M'(3))$ if $n$ is even, and $\theta(3f)=k(N'(3))=0$, where $N'$ is a characteristic submanifold of $M'$, if $n=4k+3$ (Theorem 1, IV.3.3). Therefore $M'(3)$ is normally cobordant to a homotopy projective space, whose covering sphere has normal invariant $3x$. If $n$ is even, $3x=x$; if $n$ is odd, by adding equivariantly a homotopy sphere with normal invariant $-x$ we get an involution on a homotopy sphere with normal invariant $x$.

All is left to show is that if $\Sigma\in\theta^n$ admits an involution, so does $\Sigma+\Sigma'$, for any $\Sigma'\in\theta^n(\partial\pi)$. For $n$ even this is trivial, for $n=4k+3$ this follows from V.3, and for $n=4k+1$, this will follow from the next theorem, which slightly extends results of [17]:

**Theorem.** *If $\Sigma^{4k+1}$ admits an involution $T$, then $\Sigma'=\Sigma+\Sigma_0$, where $\Sigma_0$ generates $\theta^{4k+1}(\partial\pi)$, admits an involution $T'$ such that*

$$\phi_{4i+2}(\Sigma'/T')=\phi_{4i+2}(\Sigma/T)+1 \qquad \text{for } i<k.$$

*In particular $\sigma(T',\Sigma')=\sigma(T,\Sigma)+1$.*

*Proof.* Let $Q=\Sigma/T$ and $M'=Q(3)$. Let $\eta$ be the canonical line bundle over $Q$ and $\zeta$ the pull-back of $\eta$ over $M'$. Then the normal map $f\colon M'\to Q$ induces a normal map $\bar{E}(\zeta)\to\bar{E}(\eta)$, and by connecting the components of $\bar{E}(\zeta)$ along the boundary, we obtain a normal map

$$g\colon\ U\to\bar{E}(\eta)$$

such that $\partial U=\Sigma$ and $g|\partial U$ is a homotopy equivalence, so this can be extended to a (p.l.) normal map

$$\bar{g}\colon \bar{U}\to\Sigma Q$$

where $\bar{U}=U\cup(\text{cone on }\partial U)$. $\bar{g}$ can be thought of as the result of connecting

$$(\Sigma Q)(3)\to\Sigma Q$$

by surgery. Therefore $\theta(\bar{g})=k(\bar{g})=1$, and the obstruction to making $g$ a homotopy equivalence rel. boundary is not zero.

Now $f$ is normally cobordant to a homotopy equivalence

$$f'\colon\ Q'\to Q$$

and by the normal cobordism extension property, there is a normal map

$$g'\colon\ U'\to\bar{E}(\eta)$$

normally cobordant to $g$, rel. boundary, $t$-regular at $Q$ and such that $(g')^{-1}(Q)=Q'$ and $g'|Q'=f'$. Therefore we can decompose $U'$ as the union of $\bar{E}(\eta')$ and a manifold $W$ along $\partial\bar{E}(\eta')=\tilde{Q}'$ and we can assume that $g'$ sends $\bar{E}(\eta')$ to $\bar{E}(\eta)$ as a bundle map, and $W$ into $\partial\bar{E}(\eta)=\Sigma$. Therefore $W$ is a framed cobordism between $\tilde{Q}'$ and $\partial U'=\partial U=\Sigma$, and the Kervaire invariant of $W$ is also 1, for otherwise $g$ would be normally cobordant to a homotopy equivalence rel. boundary. Therefore $\tilde{Q}'=\Sigma+\Sigma_0=\Sigma'$, and the invariants $\phi_{4i+2}$ are computed using Browder's result applied to the restrictions of the normal invariants of $Q$ and $Q'$ to $P^{4i+2}$.

This theorem gives a sharper form of Theorem II.1.B in the smooth case. Actually, Browder shows in [17] that $P^{4k+1}(m)$ is equivalent to the Brieskorn example $\Sigma_m^{4k+1}/T$ of V.4.

As an application of the proof of this theorem, we get information about the invariant $\kappa$ defined in V.3.

**Corollary 1.** *If* $\tilde{Q}^{4k+1} \in \theta^{4k+1}(\partial\pi)$, *then* $\kappa(Q(n)) = \kappa(Q)$ *for all* $n$.

**Corollary 2.** *There exists a curious involution* $(T, S^{4k+1})$ *with* $\sigma(T, S^{4k+1}) = 1$ *if, and only if, there exists a curious involution* $(T', \Sigma_0^{4k+1})$ *with* $\sigma(T', \Sigma_0^{4k+1}) = 0$.

A related result has been proved by Orlik: If $(T, \Sigma^{4k+1})$ extends to an involution with fixed points of a $\pi$-manifold whose boundary is $\Sigma$, then $\kappa(T, \Sigma) = 0$ ([67]).

## V.6 Smooth Involutions of Spheres of Low Dimension

### V.6.1 Involutions of 7-Spheres

Theorem IV.3.4 gives also the classification of smooth involutions of 5- and 6-spheres, since

$$hS(P^5) \approx hT(P^5) \approx \mathbb{Z}_4$$

$$hS(P^6) \approx hT(P^6) \approx \mathbb{Z}_4.$$

The first difference between the smooth and the p.l. cases comes in dim. 7:

**Theorem.** *There is a bijection*

$$hS(P^7) \approx \mathbb{Z}_4 \oplus \mathbb{Z}_{28} \oplus \mathbb{Z}$$

*such that the first factor gives the normal invariant, the last factor gives the Browder-Livesay invariant, and* $\theta^7 \approx \mathbb{Z}_{28}$ *acts by addition on the second factor.*

*The invariant* $\kappa: \mathbb{Z}_4 \to \mathbb{Z}_2$ *is given by reduction mod. 2, and the covering sphere* $hS(P^7) \to \theta^7$ *is given by* $(\alpha, \theta, \sigma) \to \kappa(\alpha) + 2\theta + \sigma$ *mod. 28 (where we take* $\kappa(\alpha) = 0, 1 \in \mathbb{Z}$*).*

*Proof.* Clearly $[P^7, G/0] \approx [P^7, G/PL] \approx \mathbb{Z}_4 \oplus \mathbb{Z}_2$, and the surgery obstruction is given by the $\mathbb{Z}_2$ factor, so the set of normal invariants that can be realized is $\mathbb{Z}_4$. From Theorems V.2.1 and IV.3.3 (Theorem 2′), it follows that we have an exact sequence

$$\theta^7 \to hS(P^7) \to \mathbb{Z}_4 \oplus \mathbb{Z} \to 0.$$

Now from [9], proof of Corollary 2.11 it follows that $\theta^7$ acts freely on $hS(P^7)$, so $hS(P^7) \approx \mathbb{Z}_4 \oplus \mathbb{Z}_{28} \oplus \mathbb{Z}$. (See also V.7.)

To give an explicit isomorphism we need specific representatives in each normal invariant, and for this we use the work of Montgomery and Yang [61]. They show that $hS(CP^3) \approx \mathbb{Z}$, and give a detailed description of the elements in the image of $hS(CP^3) \to hS(P^7)$. Now $hS(CP^3) \approx [CP^2, G/0] \to \mathbb{Z}_4$ is an epimorphism, so we can choose the images of the elements $0, 1, 2, 3 \in hS(CP^3)$ as representatives of the different normal invariants of $P^7$, and from [61] it follows that $\kappa$ is as it is described in the theorem (see V.3 for the definition of $\kappa$). Now choose an element $Q_\alpha^7$ within each normal invariant $\alpha$ such that $\sigma(Q_\alpha^7) = 0$ and $\tilde{Q}_\alpha^7 = \kappa(\alpha)\, \Sigma_0^7$ (there are two possibilities; for $Q_0^7$ take $P^7$), so by making $Q_\alpha^7$ correspond to $(\alpha, 0, 0)$ we get an isomorphism as described in the theorem.

The map $hS(CP^3) \to hS(P^7)$, though it was an essential element in the proof, does not have a simple expression in terms of this isomorphism.

*Added in proof.* A more detailed study of involutions of 7-spheres using spin invariants has been carried out by K.H. Mayer: Fixpunktfreie Involutionen von 7-Sphären. Math Ann. **185**, 250–258 (1970).

### V.6.2 Spheres that Admit Involutions

Let $\theta_I^n \subset \theta^n$ be the subgroup of spheres that admit (fixed point free) involutions (see V.5.2). We can compute $\theta_I^n$ for small values of $n$. For this we use the values of $\pi_n\,(G/0)$ as given in [76], and Toda's tables [77].

| $n$ | 7 | 8 | 9 | 10 | 11 | 12 | 13 | 14 | 15 |
|---|---|---|---|---|---|---|---|---|---|
| $\theta^n$ | $\mathbb{Z}_{28}$ | $\mathbb{Z}_2$ | $3\mathbb{Z}_2$ | $\mathbb{Z}_6$ | $\mathbb{Z}_{992}$ | 0 | $\mathbb{Z}_3$ | $\mathbb{Z}_2$ | $\mathbb{Z}_2 \oplus \mathbb{Z}_{8,128}$ |
| $\theta_I^n$ | $\mathbb{Z}_{28}$ | 0 | $2\mathbb{Z}_2$ | 0 | $\mathbb{Z}_{992}$ | 0 | $\mathbb{Z}_3$ | 0 | $\mathbb{Z}_2 \oplus \mathbb{Z}_{8,128}$ |

The proofs and descriptions of the examples follow:

$n=7$ was studied in V.6.1.

$n=8$: For this we show that $\pi_8^*$: $[P^8, G/0] \to \pi_8(G/0)$ is zero: Since $[CP^4, G/0] \to [CP^3, G/0] \to \pi_7(G/0) = 0$ is exact, and $[CP^3, G/0] \to [P^7, G/0]$ is onto (Lemma 2, IV.1), it follows that $[P^9, G/0] \to [P^7, G/0]$

is also onto. Now consider the diagram

$$\begin{array}{ccccc}
[P^9, G/0] & & \longrightarrow & & [P^7, G/0] \to 0 \\
& \searrow^{i_8^*} & & \nearrow^{i_7^*} & \\
& & [P^8, G/0] & & \\
& \nearrow^{j_8^*} & & \searrow^{\pi_8^*} & \\
\pi_8(G/0) & & \xrightarrow{\;0\;} & & \pi_8(G/0).
\end{array}$$

The diagonals are exact, and the lower map is zero, from the results of IV.1, and $\pi_8^*=0$, by diagram chasing.

$n=9$. We will show that $\theta_I^9$ is precisely the subgroup of spheres that bound spin manifolds ([60]). We do this in several steps:

a) *Let* $x\in[P^9, G/0]$ *be such that its restriction to* $P^7$ *is* 0. *Then* $\pi_9^*(x)=0$.

*Proof.* $S^9 \to P^9 \xrightarrow{x} G/0$ factors through

$$S^9 \to P^9/P^7 \sim S^9 \vee S^8 \to G/0.$$

The components of $S^9 \to S^9 \vee S^8$ are the map of degree 2 and the zero map $S^9 \to S^8$, since in $P^{10}/P^7$, $Sq^2\, x^8=0$. Therefore $\pi_9^*(x)\in 2\pi_9(G/0)=0$.

b) $\pi_9^*[P^9, G/0]=\operatorname{im}([CP^4, G/0]\to\pi_9(G/0))$.

*Proof.* From the commutative diagram

$$\begin{array}{ccccc}
[P^7, G/0] & \longleftarrow & [P^9, G/0] & & \\
\uparrow{\scriptstyle \text{onto}} & & \uparrow & \searrow^{\pi_9^*} & \\
& & & & \pi_9(G/0) \\
& & & \nearrow & \\
[CP^3, G/0] & \xleftarrow[\text{onto}]{} & [CP^4, G/0] & &
\end{array}$$

it follows that for every $x\in[P^9, G/0]$ there is $x'\in[P^9, G/0]$ which comes from $[CP^4, G/0]$ and has the same image in $[P^7, G/0]$. But $\pi_9^*(x)=\pi_9^*(x')$, from a), and therefore the two maps on the right have the same image, and b) is proved.

Now $\pi_9(G/0)=\pi_9(G)/\operatorname{Im} J\approx \mathbb{Z}_2\oplus\mathbb{Z}_2$, with generators (the classes of) $\nu^3$ and $\mu$ ([77]).

c) $\operatorname{im}([CP^4, G/0]\to\pi_9(G/0))$ *is generated by* $\nu^3$.

*Proof.* First we show that $\mu$ is not in the image. This is done by showing that an element in the image is represented by a sphere which bounds a spin manifold (extending an argument of R. Lee [47]). It is well known that a sphere representing $\mu$ does not bound a spin manifold ([60]).

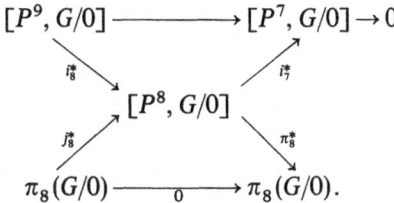

Let $f\colon M^8\to CP^4$ represent $x\in[CP^4, G/0]$, and $\xi$ be the canonical plane bundle over $CP^4$, $\xi'=f^*\xi$. Then $f$ induces a normal map $g$: $\bar{E}(\xi')\to\bar{E}(\xi)$ between the disc bundles, and its restriction to the boundary $\dot{g}\colon \dot{E}(\xi')\to S^9$ represents the image of $x$ in $\pi_9(G/0)$. Let $W$ be a normal cobordism between $g$ and a homotopy equivalence $\Sigma^9 \to S^9$, and we can combine all these in a normal map $h\colon U = \bar{E}(\xi') \cup W \to \bar{E}(\xi)$, with $\partial U = \Sigma$. But $w_i(\bar{E}(\xi)) = w_i(CP^5) = 0$ for $i = 1, 2$, and also $w_i(U) = 0$, $i = 1, 2$, since $h$ is a normal map. Therefore $\Sigma$ bounds the spin manifold $U$.

To show that $\nu^3$ is in the image, we give an explicit example (first considered by Sullivan). Consider the composition

$$S^7 \to CP^3 \to CP^3/CP^2 \sim S^6 \xrightarrow{\nu^2} G/0.$$

The map $S^7 \to S^6$ is $\eta$, since in $CP^4/CP^2$, $Sq^2\, x^3 = x^4$, and since $\eta\, \nu^2 = 0$, $CP^3 \to G/0$ extends to a mapping $CP^4 \to G/0$. We have to show that the induce map $S^9 \to G/0$ is $\nu^3$. This map factors through

$$S^9 \to CP^4/CP^2 = S^6 \cup_\eta e^8 \to G/0.$$

Taking the collapsing map $S^6 \cup_\eta e^8 \to S^8$, the composition $S^9 \to S^6 \cup_\eta e^8 \to S^8$ is zero, since in $CP^5/CP^3$ $Sq^2\, x^4 = 0$. Because we are in the stable range, this map factors through $S^9 \to S^6$, and this last map is $\nu$, mod. $2\pi_9(S^6)$, because in $CP^5/CP^2$, $Sq^4\, x^3 = x^5$. Therefore the composition $S^9 \to G/0$ is $\nu^3$. The result now follows from b), c) and V.4.2.

*$n = 10$.* If $\Sigma \in \theta^{10}$ admits an involution, then it bounds a spin manifold (R. Lee [47]). This follows as before, since $\Sigma$ bounds the mapping cylinder of $\Sigma \to \Sigma/T$, whose Stiefel-Whitney classes are those of $P^{11}$. Of the elements of $\theta^{10} \approx \mathbb{Z}_6$, only $\eta \circ \mu$ is of order 2 and could admit an involution, but it does not bound a spin manifold.

*$n = 11, 12, 13$* follow directly from V.3, since $\theta^{11} = \theta^{11}(\partial\pi)$ and $\theta^{13}$ is of odd order.

*$n = 14$.* Let $X$ be an $h$-space such that $\Omega X = G/0$, which exists because $G/0$ is an infinite loop space ([7]). From IV.1 we have the diagram

$$\begin{array}{ccccc} [\Sigma P^{14}, X] & \longrightarrow & \pi_{15}(X) & \longrightarrow & [P^{15}, X] \\ \wr\wr & & \wr\wr & & \\ [P^{14}, G/0] & \longrightarrow & \pi_{14}(G/0) & & \end{array}$$

Therefore, to show that an element in $\pi_{14}(G/0)$ is not in the image of $\pi_{14}^*$, is is enough to show that its image in $[P^{15}, X]$ is not zero. Now $\pi_{14}(G/0) \approx \pi_{14}(G) \approx \mathbb{Z}_2 \oplus \mathbb{Z}_2$ generated by $\sigma^2$ and $\kappa$. Since $\sigma^2$ has Kervaire invariant 1, it does not represent a sphere (and in fact is not in the image of $\pi_{14}^*$, because its image in $\pi_{14}(G/PL)$ is not, IV.2), so we have only to check $\kappa$. Considering the diagram

$$\begin{array}{ccccc} S^{15} & & & & \\ \downarrow & \searrow^{\times 2} & & & \\ P^{15} & \longrightarrow & S^{15} & \xrightarrow{\kappa} & X \end{array}$$

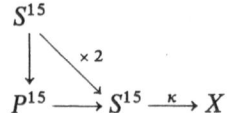

it follows that $P^{15} \to X$ extends to $P^{16}$

$$\begin{array}{ccccc} S^{16} & \longrightarrow & P^{16} & \longrightarrow & P^{16}/P^{14} \\ & & \cup & & \cup \quad \searrow \\ & & P^{15} & \longrightarrow & S^{15} \longrightarrow X \end{array}$$

$S^{16} \to P^{16}/P^{14} \to S^{16}$ is zero, so $S^{16} \to X$ factors through $S^{16} \to S^{15} \xrightarrow{\kappa} X$. But the map $S^{16} \to S^{15}$ is $\eta$, since in $P^{17}/P^{14}$, $Sq^2\, x^{15} = x^{17}$, so the composition is $\eta \circ \kappa \neq 0$. Since this holds for any extension to $P^{16}$ of $P^{15} \to X$, this map is not trivial and the result is proved.

*$n=15$*. Take the map $CP^7 \to S^{14} \xrightarrow{\varepsilon} G/0$. (This gives a normal map corresponding to $CP^7 \# \Sigma^{14}$, where $\Sigma^{14}$ is the exotic 14-sphere, and was first considered by Bredon [9].) The composition $S^{15} \to CP^7 \to G/0$ is $\eta \circ \varepsilon$, which is the normal invariant of a 15-sphere that does not bound a $\pi$-manifold. Since all other spheres can be obtained from this sphere and the standard one by adding elements of $\theta^{15}(\partial\pi) \approx \mathbb{Z}_{8,128}$, the result follows from V.3.

The example in dimension 9 shows that $G/0$ does not satisfy condition (!) of IV.1, and the example in dimension 15, which is p.l. equivalent to the antipodal map, shows that $PL/0$ does not satisfy it either.

Clearly the methods used in computing $\theta_I^n$ for $n \leqq 15$ can be generalized to obtain systematic results about spheres which do or do not admit involutions.

It should be mentioned here that in all dimensions bigger than 7 there are p.l. involutions that cannot be smoothed ([31]), in other words the maps $hS(P^n) \to hT(P^n)$ and $[P^n, G/0] \to [P^n, G/PL]$ are not onto for $n \geqq 8$. For example, only 4 of the 8 involutions of $S^8$, and only 8 of the 16 involutions of $S^9$ can be smoothed.

## V.7 Action of $\theta^n(\partial\pi)$ (After Browder)

The only element of the cyclic group $\theta^{4k-1}(\partial\pi)$ that can possibly act trivially on a homotopy projective space $Q^{4k-1}$ is the element of order 2, for the action of any other element can be detected by the double covering. To decide whether the element of order 2 acts trivially or not is a subtle question. For completeness we include (without proof) a result of Browder that settles the question in a good number of cases.

Let $d_k = \text{denominator}\left(\frac{B_k}{4k}\right)$, where $B_k$ is the $k$-th Bernoulli number, $j_k$ = order of the image of $J$: $\pi_{4k-1}(SO) \to \pi_{4k-1}(G) = \pi^S_{4k-1}$, and $q$: $\Sigma P^{4k-1} \to S^{4k}$ the map of degree 1.

**Theorem** (Browder). a) *If* $j_k = d_k$, then $q^*\colon J(S^{4k}) \to J(\Sigma P^{4k-1})$ *is a monomorphism.*

b) *If* $q^*\colon J(S^{4k}) \to J(\Sigma P^{4k-1})$ *is a monomorphism then* $\theta^{4k-1}(\partial\pi)$ *acts freely on any homotopy projective space* $Q^{4k-1}$.

*Remarks.* It has been shown by Adams [1], that $j_k = \varepsilon\, d_k$, where $\varepsilon = 1$ or 2, and that $\varepsilon = 1$ for $k$ odd. The general result for $k$ even is still not settled, but Mahowald has proved that $\varepsilon = 1$ if $k$ is a power of 2, and also for a large number of small values of $k$ ([56]).

## V.8 How to Prove $\sigma = \tau$

The more precise result $\sigma = \tau$ (compare with IV.4 and V.2.3) can be proved, following [33], [34] and [36], in an elementary way. We give here a sketch of this proof.

*The index.* Let $M^{4k}$ be an oriented compact manifold, with or without boundary, and let $T\colon M \to M$ be an orientation preserving involution, possibly with fixed points. Define the bilinear form $B_T$ on $H_{2k}(M)$ by

$$B_T(x, y) = x \cdot T_* y$$

and $I(T, M) = \text{Index } B_T$. Let Fix $T$ denote the set of fixed points of $M$ under $T$. Then the following properties hold:

A) *If Fix* $T = \phi$, *then* $I(T, M) = 2 \text{ Index } (M/T) - \text{Index } M$. *If further* $\partial M = \phi$, *then* $I(T, M) = 0$.

*Proof.* Let $H = H_{2k}(M; \mathbb{R})$, $A = (1 + T_*) H$, $B = (1 - T_*) H$. Then:

a) $H = A \oplus B$, and the summands are orthogonal both for the intersection form and $B_T$.

b) On $A$, $B_T(x, y) = x \cdot y$, while on $B$, $B_T(x, y) = -x \cdot y$.

c) If $p\colon M \to M/T$ is the projection, $p_*|A\colon A \to H_{2k}(M/T; R)$ is an isomorphism, and, $p_*|B = 0$. Furthermore $p_*(x) \cdot p_*(y) = x \cdot y + x \cdot Ty = 2x \cdot y$ for $x, y \in A$, so Index $(B_T|A) = \text{Index } (M/T)$.

Therefore

$$I(T, M) = \text{Index } (B_T|A) + \text{Index } (B_T|B)$$

$$\text{Index } (M) = \text{Index } (B_T|A) - \text{Index } (B_T|B)$$

and if $\partial M = \phi$, Index $M = 2$ Index $(M/T)$, by the Dold construction (see below), so A) follows.

B) *Novikov's addition lemma. If* $M^{4k} = M_0^{4k} \cup M_1^{4k}$, $M_0$, $M_1$ *invariant compact submanifolds, with* $M_0 \cap M_1 =$ *union of components of* $\partial M_i$, $i = 1, 2$, *then*

$$I(T, M) = I(T|M_0, M_0) + I(T|M_1, M_1).$$

For a proof see [4], p. 588.

We will say that Fix $T$ is regular if

a) Fix $T$ is a submanifold of the interior of $M$.

b) Fix $T$ has a linear normal bundle in $M$, which embeds in $M$ as a tubular neighborhood of Fix $T$ in such a way that $T$ corresponds to the antipodal map in each fibre.

It is known that b) always holds in the smooth case, but not always in the p.l. case.

C) *If* $\partial M=\phi$, *and* Fix $T$ *is regular and admits a non-zero normal section, then* $I(T, M)=0$.

*Proof* ([36]). Let $U$ be the normal closed disc bundle of Fix $T$, and $Y=M-U$. $U$ is invariant and $I(T|U, U)=0$, since in $U$ all intersection numbers are 0, because cycles can be pushed away from each other using the section. Now let $f$: Fix $T\to\partial U=\partial Y$ be the normalized section. Define $\psi\colon\partial Y\to\partial Y$ to be the reflection on $f(x)^{\perp}$ on the fiber over $x$. $\psi$ is orientation reversing and equivariant, so we can form the involution $(T', Y\cup_{\psi} Y)$. From A) and B), $0=I(T', Y\cup_{\psi} Y)=2I(T|Y, Y)$, and finally $I(T, M)=I(T|U, U)+I(T|Y, Y)=0$.

*The Dold construction.* Let $(T, N)$ be an orientation preserving fixed point free involution of a closed manifold $N$, and $W$ a characteristic submanifold for $(T, N)$. The Dold construction $\mathscr{D}=\mathscr{D}(T, N, W)$ is a manifold with an involution $\mathscr{T}$, such that $\partial\mathscr{D}=(T, N)-2(N/T)$, where the involution on $2(N/T)$ interchanges the two copies, and Fix $\mathscr{T}=W/T$ is regular and admits a non-zero normal section. It is constructed as follows: If $W=A\cap TA$, then $N/T=A\cup E$, where $E$ is the mapping cylinder of $W\to W/T$. On $E$ define an involution $T'$ by $T'(w, t)=(T(w), t)$. On $W=\partial E$, $T'=T$. (Thinking of $E$ as line bundle over $W/T$, $T'$ is the antipodal map on each fibre.) Form $\mathscr{D}=(N/T\times I)\cup_{(T', E\times 1)}(N/T\times I)^*$. (See II.2.1.)

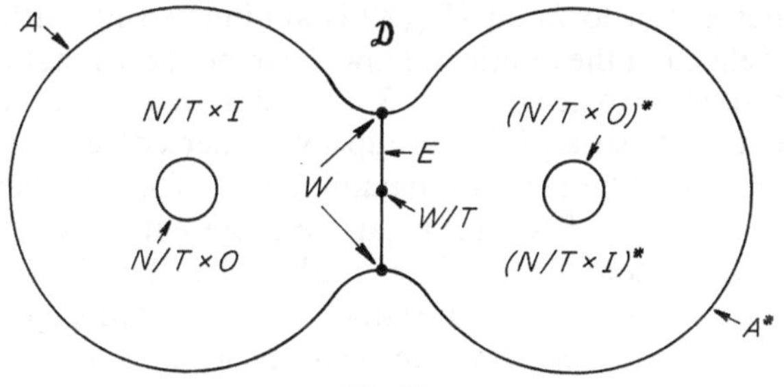


Fig. 11

Clearly $\mathscr{D}$ has the stated properties, and this proves

D) *Given $(T, N)$, an orientation preserving fixed point free involution of a closed manifold $N$, there is an equivariant cobordism between $2(N/T)$ and $(T, N)$, whose fixed point set is regular and has a non-zero normal section.*

*If* $\dim N = 4k+3$, *and* $(T', M)$ *is any such cobordism, then* $\alpha(T, N) = I(T', M)$ *is well defined.*

The last part follows from B) and C): If $(T_1', M_1)$ and $(T_2', M_2)$ are two such cobordisms. $(T_1', M_1) \cup_{\partial M_1} -(T_2', M_2)$ satisfies the hypotheses of C).

E) $\alpha(T, N) = 8\sigma(T, N)$. (See definition at the end of Chapter I.)

This can be shown for the Dold construction as follows: Consider the Mayer-Vietoris sequences of $(N, A, TA)$ and $(\mathscr{D}, B, B^*)$, where $B$ denotes $N/T \times I$, and all homology groups are with real coefficients:

$$\begin{array}{ccccccc}
 & & & & H_{2k}(E, W) & \xrightarrow{\approx} & H_{2k}(N/T, A) \\
 & & & & \downarrow & & \downarrow \\
 & & H_{2k}(N) & \xrightarrow{\Delta} & H_{2k-1}(W) & \longrightarrow & H_{2k-1}(A) \oplus H_{2k-1}(TA) \\
 & & \downarrow & & \downarrow & & \downarrow \\
\cdots & \xrightarrow{i_{1*}+i_{2*}} & H_{2k}(\mathscr{D}) & \xrightarrow{\Delta'} & H_{2k-1}(E) & \longrightarrow & H_{2k-1}(B) \oplus H_{2k-1}(B^*) \\
 & & & & \| & & \\
 & & & & H_{2k-1}(W/T) & &
\end{array}$$

It follows from the arguments in the proof of A) and diagram chasing that

$$\frac{H_{2k}(\mathscr{D})}{\operatorname{im}(i_{1*} + i_{2*})} \approx \operatorname{im} \Delta' \approx \frac{K_{2k-1}}{\operatorname{im} \Delta}.$$

The intersection form on $H_{2k}(\mathscr{D})$ is annihilated by $\operatorname{im}(i_{1*} + i_{2*})$, so $B_T$ is well defined in the quotient. Now it can be shown that this bilinear form corresponds to $x \cdot T_* y$ on $K_{2k-1}/\operatorname{im} \Delta$ under the isomorphism by representing a basis of this group by immersed submanifolds $V_i$ of $W \times 0$ which bound immersed submanifolds $Z_i$ of $A \times 0$. Then the corresponding basis of the first group can be represented by the manifolds $Z_i \cup Y_i \times I \cup Z_i^*$, where $Y_i \times I$ is immersed in $(E \times I) \cup (E \times I)^*$ by joining $Y_i$ and $Y_i^*$ linearly. All these immersions can be moved slightly inside $\mathscr{D}$, so that all the intersections and self-intersections outside $W/T$ are removed.

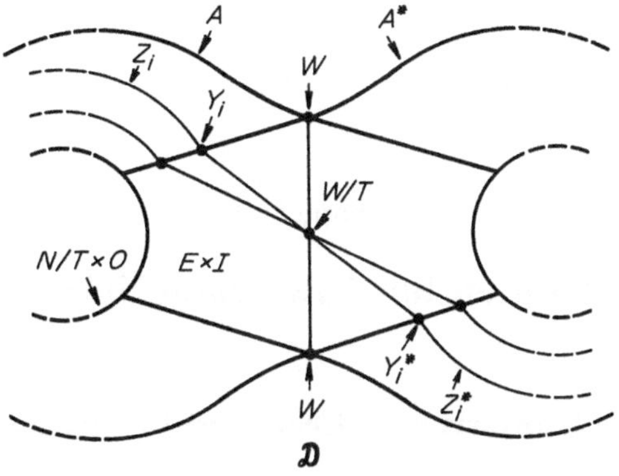


Fig. 12

Thus intersection numbers in $\mathscr{D}$ correspond to intersection numbers in $W/T$, and therefore in $W$. Notice also that in $H_{2k}(\mathscr{D})$, $x \cdot T_* y = -x \cdot y$.

**Theorem.** *If $f:(T', N') \to (T, N)$ is an equivariant homotopy equivalence, then $\sigma(f)=\tau(f)$.*

*Proof.* It is clear from the definition of $\tau$ (V.2.3) and A), B) and E) above, that

$$8\tau(f)=\alpha(T', N')-\alpha(T_0, N_0)=8\sigma(T', N')-8\sigma(T_0, N_0)=8\sigma(f).$$

Chapter VI

# Codimension 2 Invariant Spheres

## VI.1 Invariant vs. Characteristic Spheres

**Definition.** An *invariant q-sphere* for $(T, \Sigma^n)$ is an embedded (locally flat in the p.l. case) homotopy sphere $\Sigma^q \subset \Sigma^n$ which is invariant under $T$. An invariant $\Sigma^q$ for $(T, \Sigma^n)$ is called *characteristic* if there is an equivariant homotopy equivalence $f: (T, \Sigma^n) \to (a, S^n)$, ($a$ = antipodal map) such that $f$ is $t$-regular at $S^q \subset S^n$ and $f^{-1}(S^q) = \Sigma^q$.

Clearly an invariant $\Sigma^q$ for $(T, \Sigma^n)$ corresponds to an embedded homotopy projective space $Q^q = \Sigma^q/T \subset \Sigma^n/T = Q^n$ such that the inclusion induces an isomorphism of fundamental groups, and $Q^q$ is characteristic if there is a homotopy equivalence $Q^n \to P^n$ such that the $t$-regular inverse image of $P^q$ is $Q^q$.

The difference between invariant and characteristic spheres reflects the two possible ways of obtaining them:

1) Given $Q^n$, choose any $Q^q$ and try to change the natural map $Q^q \to Q^n$ into an embedding. In the smooth case this works well in the metastable range $2n \geqq 3(q+1)$ ([28]) and in the p.l. case if $n \geqq q+3$, by the Casson-Haefliger-Sullivan theorem (see, for example, [29]).

2) Given $Q^n$, try to make the homotopy equivalence $f: Q^n \to P^n$ $h$-regular at $P^q$, in which case $Q^q = f^{-1}(P^q)$. A main difference with the first approach is that in this case the normal invariant of $Q^q$ is determined to be the restriction of the normal invariant of $Q^n$ to $P^q$ (III.1.3). Therefore if the restricted normal invariant cannot be realized, $(T, \Sigma^n)$ does not admit a characteristic $q$-sphere, and this can happen with arbitrary codimension. For $n \geqq q+3$ it can be shown, using the methods of Browder's embedding theorem [15], that $(T, \Sigma^n)$ admits a characteristic $q$-sphere if, and only if, the restriction of the normal invariant to $P^q$ can be realized by a $Q^q$. For $n = q+2$, $n \neq 4$, the same result follows from Theorem 1, VI.3. On the other hand, we know that for $n = q+1$ this is not always the case (II.4, III.3.2 and V.2.1). In particular $(T, \Sigma^n)$ admits a characteristic $(4k+1)$-sphere for all $4k+1 < n$. Also, if $2n > 3(q+1)$ in the smooth case, and if $n \geqq q+3$ in the p.l. case, an invariant $\Sigma^q$ for $(T, \Sigma^n)$ is characteristic

if, and only if, $\Sigma^q/T$ has the right normal invariant. Combining this with the following theorem, it follows that this is true in the p.l. case without restrictions.

For $q \geqq n-2$ the two cases merge:

**Theorem.** *If $q=n-1$ or $q=n-2$, then any invariant $\Sigma^q$ for $(T, \Sigma^n)$ is characteristic.*

*Proof.* We will work with the quotient spaces $Q^q = \Sigma^q/T$ and $Q^n = \Sigma^n/T$.

If $q=n-1$, choose a homotopy equivalence $g\colon Q^{n-1} \to P^{n-1}$. Since the normal bundle $\nu$ of $Q^{n-1}$ in $Q^n$ must be the pull-back of the Hopf bundle $\eta$ over $P^{n-1}$, $g$ can be covered by a bundle map $b\colon E(\nu) \to E(\eta)$, which by the Thom-Pontryagin construction induces a map $f\colon Q^n \to T(\eta) = P^n$ which is $t$-regular at $P^{n-1}$ with inverse image $Q^{n-1}$ and has degree $\pm 1$ and is therefore a homotopy equivalence (IV.3.1).

If $q=n-2$, again choose a homotopy equivalence $g\colon Q^{n-2} \to P^{n-2}$. In the smooth case the normal bundle $\nu$ of $Q^{n-2}$ in $Q^n$ must have the same Stiefel-Whitney classes as the pull-back of $\eta \oplus \eta$: $w_1 = 0$, $w_2 \neq 0$. Therefore $\nu$ is orientable and it is classified by its Euler-class, which in this case can be identified with $w_2$, so indeed $\nu = g^*(\eta \oplus \eta)$. In the p.l. case it follows from [79] that $Q^{n-2} \subset Q^n$ has a normal bundle, and since p.l. and linear bundles coincide in dimension 2, the above argument can again be applied.

So in any case $g$ can be covered by a bundle map $b\colon \nu \to \eta \oplus \eta$. Identifying $E(\nu)$ and $E(\eta \oplus \eta)$ with tubular neighborhoods $U$ and $V$ of $Q^{n-2}$ in $Q^n$ and $P^{n-2}$ in $P^n$, respectively, $b$ induces a map $g'\colon \bar{U} \to \bar{V}$. We would like to extend $g' | \partial \bar{U}\colon \partial \bar{U} \to \partial \bar{V}$ to a map $Q^n - U \to P^n - V \simeq S^1$, and since $S^1 = K(\mathbb{Z}, 1)$ this can be done if, and only if, the cohomology class $g'^*(i) \in H^1(\partial U)$ is the restriction of a class in $H^1(Q^n - U)$. From the cohomology exact sequence

$$H^1(Q^n - U) \xrightarrow{i^*} H^1(\partial \bar{U}) \to H^2(Q^n - U, \partial \bar{U})$$

it follows that $i^*$ is onto if $n \neq 3$, since $H^2(Q^n - U, \partial \bar{U}) \approx H^2(Q^n, Q^{n-2}) = 0$, because $Q^{n-2} \subset Q^n$ induces an isomorphism of fundamental groups and therefore looks homologically as $P^{n-2} \subset P^n$.

For the case $n=3$, an argument analogous to that given in [41] shows again that $g'^*(i)$ is in the image of $i^*$. So in all cases $g'$ can be extended to a map $f\colon Q^n \to P^n$ that is $t$-regular at $P^{n-2}$, $f^{-1}(P^{n-2}) = Q^{n-2}$ and has degree $\pm 1$, and is therefore a homotopy equivalence.

The result is best possible: choose $Q^6$ with $\phi_2(Q^6) \neq 0$. Then we can embed $P^3 \subset Q^6$, but it cannot be characteristic because it has the wrong normal invariant (in fact, $Q^6$ does not admit any characteristic $Q^3$).

This theorem has been proved independently by Orlik and Rourke, who also apply it to obtain a version of Theorem 1, VI.3.

## VI.2 Applications

### VI.2.1 A Non-embedding Result

Consider $Q^5$ with $\phi_2(Q^5) \neq 0$. Then the natural map $P^3 \to Q^5$ is not homotopic to a locally flat embedding. This example barely fails to satisfy the hypotheses of the various known embedding theorems. It is still possible that a non-locally-flat embedding can be found as in the examples of [43]. Notice that the other possible map $P^3 \to Q^5$, the null-homotopic one, is homotopic to a nice embedding because $P^3 \subset R^5$.

### VI.2.2 Some Brieskorn-Hirzebruch Involutions

Consider the submanifold of $\mathbb{C}^{m+1}$, $m \geqq 3$ given by the equations

$$z_0^{a_0} + \cdots + z_m^{a_m} = 0$$

$$z_0 \bar{z}_0 + \cdots + z_m \bar{z}_m = 1$$

where the numbers $a_0, \ldots, a_m$ are all odd $\geqq 3$, and for at least two values of $i$, $a_i$ is relatively prime to all $a_j$, $j \neq i$. Then this manifold is a homotopy sphere ([10, 32]), which will be denoted by $\Sigma_a^{2m-1} = \Sigma_{(a_0, \ldots, a_m)}^{2m-1}$. The involution $(T, \Sigma_a)$ given by $z_i \to -z_i$ has been considered by Hirzebruch [33].

**Theorem.** *$\Sigma_a/T$ has the same normal invariant as $P^{2m-1}$.*

*Proof.* By construction $(T, \Sigma_a) \subset (a, S^{2m+1})$, the antipodal involution. By VI.1 $(T, \Sigma_a)$ is characteristic and therefore it has the same normal invariant as $(a, S^{2m-1})$.

Notice that this is an equivariant version of the well-known fact that if $\Sigma^n \subset S^{n+2}$, then $\Sigma^n$ bounds a $\pi$-manifold.

If $m = 2k$, then $(T, \Sigma_a^{4k-1})$ must be diffeomorphic with one of the involutions obtained from the antipodal map by construction II.4. Since the invariant $\kappa$ (V.3) depends only on the normal invariant, it follows that $\kappa = 0$ for all involutions of this type, and there is the following relation between $\sigma(T, \Sigma_a^{4k-1})$ and the exponents $a_0, \ldots, a_{2k}$: It is shown in [32] that $\Sigma_a^{4k-1} = (-1)^k \frac{1}{8} \tau(a) \Sigma_0^{4k-1}$, where $\Sigma_0^{4k-1}$ is the generator of $\theta^{4k-1}(\partial\pi)$ and $\tau(a)=\tau(a_0, \ldots, a_{2k})$ is a function of the exponents described there. Therefore

$$\sigma(T, \Sigma_a^{4k-1}) \equiv \tfrac{1}{8}\tau(a_0, \ldots, a_{2k}) \quad \text{mod. } 2.$$

In particular, from computations done in [33],

$$\sigma(T, \Sigma^{4k-1}_{(5,17,3,\ldots,3)}) \neq 0.$$

The problem of determining $\sigma(T, \Sigma_a^{4k-1})$ in terms of the exponents is still open.

If $m=2k+1$, we conclude that $(T, \Sigma_a^{4k+1})$ is diffeomorphic to $(a, S^{4k+1})$ or to $(a, S^{4k+1}) \# \Sigma_0^{4k+1}$, where $\Sigma_0^{4k+1}$ is the Kervaire sphere, so all these involutions are p.l. equivalent, and in some cases even diffeomorphic, to the antipodal map. In particular $\sigma(T, \Sigma_a^{4k+1})=0$ always. (Compare V.4.)

A corollary is that a Brieskorn sphere $\Sigma^{4k+1}_{(a_0, \ldots, a_{2k+1})}$ with $a_i$ odd is diffeomorphic to the standard sphere, a well known fact ([10]).

## VI.3 Knotted and Unknotted Codimension 2 Invariant Spheres

**Definiton.** We introduce an invariant $\rho(T, \Sigma^n)$,

$$\rho(T, \Sigma^n) \in \begin{cases} \mathbb{Z}_2 & n \not\equiv 3 \quad \text{mod. } 4 \\ 0 & n \equiv 3 \quad \text{mod. } 4 \end{cases}$$

as follows:

$$\rho(T, \Sigma^{4k+1}) = \sigma(T, \Sigma^{4k+1})$$

$$\rho(T, \Sigma^{2k}) = \sigma(T, S^{2k-1}) \quad \text{mod. } 2$$

where $(T, S^{2k-1})$ is any desuspension of $(T, \Sigma^{2k})$.

By I.1.3, II.3 and IV.4.2 $\rho$ is well defined, except possibly for $n=4$.

**Theorem 1.** *If $n \equiv 0$ or 3 mod. 4, $n \geqq 7$, then $(T, \Sigma^n)$ admits an invariant $\Sigma^{n-2}$. If $n \equiv 1$ or 2 mod. 4, $n \geqq 7$, then $(T, \Sigma^n)$ admits an invariant $\Sigma^{n-2}$ if, and only if, $\rho(T, \Sigma^n)=0$.*

**Theorem 2.** *If $n \geqq 7$ the following conditions are equivalent:*

(i) *$(T, \Sigma^n)$ admits an unknotted invariant $S^{n-2}$.*

(ii) *$(T, \Sigma^n)$ is a double suspension.*

(iii) $\sigma(T, \Sigma^n) = \rho(T, \Sigma^n) = 0$.

*Remarks.* Thus $\rho$ appears as the only obstruction to finding an invariant $\Sigma^{n-2}$, except for the case $n=4k$, when it isn't even an obstruction. The proof of this case is considerably more elaborate than the other cases and is given at the end of the section.

*Proofs.* Except for the case $n=4k$, Theorem 1 follows from II.7, VI.1 and the results of Chapter IV, since $\rho=\phi_{4k-2}$ for $n=4k+1$ and $4k+2$.

As for Theorem 2, the equivalence of (ii) and (iii) follows directly from I.1.3 and the definition of $\rho$, while it is clear that (ii)$\Rightarrow$(i). We now show (i)$\Rightarrow$(ii).

Let $S^{n-2}$ be an unknotted invariant sphere for $(T, \Sigma^n)$. We look at the quotients $Q^{n-2}=S^{n-2}/T$ and $Q^n=\Sigma^n/T$, and let $\nu$ be the normal bundle of $Q^{n-2}$ in $Q^n$. It was shown in VI.1 that there is a homotopy equivalence $f\colon Q^n\to P^n$ that is $t$-regular at $P^{n-2}$, $f^{-1}(P^{n-2})=Q^{n-2}$, and $\nu=f^*(\eta\oplus\eta)$. Identify $E(\nu)$ and $E(\eta\oplus\eta)$ with tubular neighborhoods of $Q^{n-2}$ and $P^{n-2}$, respectively, and we can assume that with respect to this identification $f$ restricted to the tubular neighborhood is a bundle map, and is, therefore, $t$-regular at $E(\eta)\subset E(\eta\oplus\eta)$ and $f^{-1}(E(\eta))=E(f^*(\eta))$. Let $A$ and $B$ be the complements of the tubular neighborhoods, $f'=f|A\colon A\to B\simeq S^1$, and since $S^{n-2}$ is unknotted in $\Sigma^n$, $A$ has also the homotopy type of $S^1$. Now $\dot{E}(\eta)\subset\partial B$ is a fibre of $\partial B\to S^1$, and therefore $\dot{E}(f^*\eta)$ is a fibre of $\partial A\to S^1$. By the Fibering Theorem ([20]), $\dot{E}(f^*\eta)$ is the boundary of a fiber $D^{n-1}$ for $A\to S^1$, and $D^{n-1}$ must be an $(n-1)$-disc, since $A\simeq S^1$. Therefore $Q^{n-2}\subset Q^{n-1}=E(f^*\eta)\cup D^{n-1}\subset Q^n$ and $(T, \Sigma^n)$ desuspends twice.

For the realization of the invariant $\rho$, the Realization Theorem II.1 and suspension show that it can always be realized *in the p.l. case.* So one obtains assorted examples of involutions that do not admit codimension 2 invariant spheres, that only admit knotted ones, etc. For $n$ odd these examples can be taken to be smooth, but for $n$ even the existence of smooth examples with $\rho\neq 0$ is equivalent to the existence of curious involutions (V.5.2, Corollary 2) so we have examples only in a few dimensions.

For $n\not\equiv 1$ mod. 4, an invariant codimension 2 sphere must be diffeomorphic to the standard sphere, while for $n=4k+1$, the Brieskorn examples mentioned above show that it can be any element of $\theta^{4k-1}(\partial\pi)$.

We can also consider the uniqueness problem: let $\Sigma_0^{n-2}$ and $\Sigma_1^{n-2}$ be two invariant spheres for $(T, \Sigma^n)$. Then, by VI.1, the normal invariants of $\Sigma_0^{n-2}/T$ and $\Sigma_1^{n-2}/T$ must be equal, and the examples show again that this is all we can say, so we know what the possibilities are.

If $\Sigma_0^{n-2}$ and $\Sigma_1^{n-2}$ are unknotted, then we can apply I.1.4 to obtain concordance results: If $n$ is even, then $\Sigma_0^{n-2}$ and $\Sigma_1^{n-2}$ are equivariantly concordant in $\Sigma^n \times I$ (by II.3 and IV.4.2). If $n=4k+1$ there is an obstruction that lies in $\mathbb{Z}$, and can be identified with one half of the difference between the Browder-Livesay invariants of $(T|\Sigma_0^{n-2}, \Sigma_0^{n-2})$ and $(T|\Sigma_1^{n-2}, \Sigma_1^{n-2})$ by IV.4.2. If $n=4k+3$ there is an obstruction that lies in $\mathbb{Z}_2$. These obstructions can always be realized. Except for the last case, $\Sigma_0^{n-2}$ and $\Sigma_1^{n-2}$ are equivariantly concordant if, and only if, they are equivalent as involutions, and for that case we obtain as in II.3:

**Corollary.** *There are two locally flat p.l. concordance classes of locally flat p.l. embeddings of $P^{4k+1}$ in $P^{4k+3}$ such that the complement has the homotopy type of $S^1$. In particular there are two such embeddings that are not ambient isotopic.*

Again, there are a few smooth examples. This contrasts sharply with Levine's unknotting theorem ([48], [20]). Cf. also VI.4.

The rest of this section is devoted to the proof of Theorem 1 in the case $n=4k$, $k>1$.

We first identify the invariant $\rho$ with another surgery invariant. Let $U$ be a tubular neighborhood of $P^{4k-2}$ in $P^{4k}$ and $X=P^{4k}-U$. $X$ can be identified with the total space of the non-orientable $(4k-1)$-disc bundle over $S^1$, where the 0-section corresponds to the standard $S^1=P^1$ that links $P^{4k-2}$. Therefore $X$ is non-orientable and $\pi_1(X)=\mathbb{Z}$. Now let $f\colon Q^{4k}\to P^{4k}$ be a homotopy equivalence, $t$-regular at $P^{4k-2}$, $N=f^{-1}(P^{4k-2})$ and $g=f|N$. Since $\theta(g)=\theta(f)=0$ (Theorem 1, IV.3.3), $g$ is normally cobordant to a homotopy equivalence $g'\colon Q^{4k-2}\to P^{4k-2}$, and by the normal cobordism extension lemma, $f$ is normally cobordant to a normal map $f'\colon M^{4k}\to P^{4k}$, $t$-regular at $P^{4k-2}$, such that $f'^{-1}(P^{4k-2})=Q^{4k-2}$ and $f'|Q^{4k-2}=g'$. We can further assume that $f'$ sends a tubular neighborhood $U'$ of $Q^{4k-2}$ in $M$ onto $U$ by a bundle map and $X'=M-U'$ onto $X$. Let $h=f'|X'\colon X'\to X$, then $h$ is a normal map, $h|\partial X$ is a homotopy equivalence, so we can define

$$\rho'(Q^{4k})=\theta(h)\in L_{4k}(\mathbb{Z},-).$$

To show $\rho'$ is well defined, let $f_i'\colon M_i\to P^{4k}$, $i=0,1$, be two normal maps with the same properties as $f'$, and $F\colon W^{4k+1}\to P^{4k}$ a normal cobordism between them. We can assume that $F$ is $t$-regular at $P^{4k-2}$ and that $V=F^{-1}(P^{4k-2})$ is an $h$-cobordism between the $Q_i=f_i'^{-1}(P^{4k-2})$, using the fact that $L_{4k-1}(\mathbb{Z}_2,-)=0$ and the normal cobordism extension lemma. We can further assume that $F$ sends a tubular neighborhood $U''$ of $V$ onto $U$ and $Y=W-U''$ onto $X$. Therefore $F|Y$ is a normal cobordism between $h_1$ and $h_0$, rel. boundary, so $\theta(h_0)=\theta(h_1)$.

The surgery obstruction $\theta(h)$ can be identified with an index mod. 2, using the methods of [14] (see also [82]) as follows: $h|\partial X'$ is a homotopy equivalence which we can assume $h$-regular at a fibre $S^{4k-2}$ of $\partial X\to S^1$, since we can assume $f'|U'$ is a bundle map, and make $h$ $t$-regular at the fibre $D^{4k-1}$ of $X \to S^1$ whose boundary is $S^{4k-2}$. If $W = h^{-1}(D)$, the surgery obstruction of $h|W$ rel. boundary is 0, so we can again assume $W = D'$ and $h|D'$: $D' \to D$ a homotopy equivalence. If we cut $X$ and $X'$ along $D$ and $D'$ respectively, obtaining $\hat{X}$ and $\hat{X}'$ (i.e. $\hat{X} = X -$ collar neighborhood of $D$, etc.), we get a normal map $\hat{h}$: $\hat{X}' \to \hat{X}$ such that $\hat{h}|\partial\hat{X}'$ is a homotopy equivalence. Since $\hat{X}$ is a $4k$-disc, we have $\theta(\hat{h}) = \sigma(\hat{X}') \in \mathbb{Z}$.

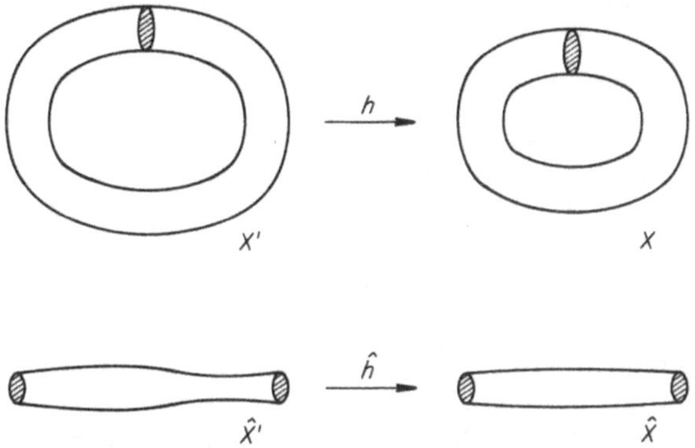

Fig. 13

Now we show that $\theta(h)$ can be identified with the mod. 2 class of $\theta(\hat{h})$:

a) First of all, $\theta(\hat{h})$ mod. 2 is an invariant of the normal cobordism class of $h$, rel. boundary. For if $h_1$ is another normal map in this class, with the same properties as $h$, and $H$: $Y \to X$ is a normal cobordism, rel. boundary, between $h_1$ and $h$, we can assume that $H$ is $t$-regular at $D$, $V = H^{-1}(D)$, and $H^{-1}(S)$ is an $h$-cobordism between $\partial D_1'$ and $\partial D'$. Cutting $Y$ along $V$ we obtain a normal cobordism $\hat{H}$: $\hat{Y} \to \hat{X}$ between $\hat{h}_1$ and $\hat{h}$.

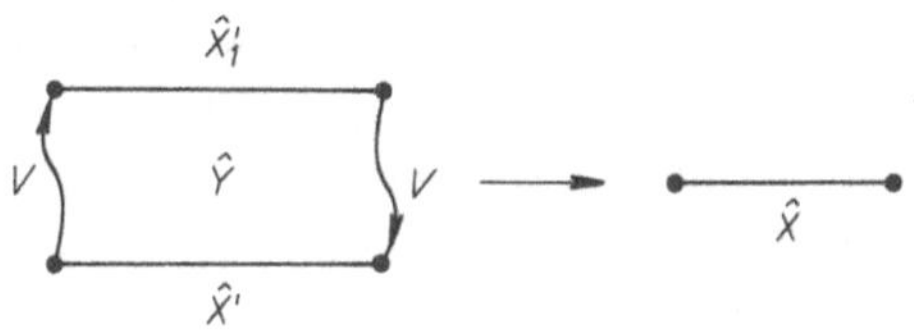

Fig. 14

Now, $\hat{Y}$ can be thought of as a cobordism, rel. boundary, between $\hat{X}_1'$ and $V \cup \hat{X}' \cup V$ ($V$ gets the same orientation twice because $Y$ is non-orientable). Therefore

$$\sigma(\hat{X}_1') = \sigma(\hat{X}') + 2\sigma(V)$$

and $\sigma(\hat{X}')$ is well defined mod. 2.

b) If $\theta(\hat{h})$ is even, then $h$ is normally cobordant to a homotopy equivalence. Because in that case we can find $V$ as above such that $2\sigma(V) = -\sigma(\hat{X}')$, and by the normal cobordism extension lemma, find a cobordism $H$: $Y \to X$ between $h$ and $h_1$ with $H^{-1}(D) = V$. Then $\sigma(\hat{X}_1') = 0$ and we can perform surgery on $\hat{X}_1'$ to obtain a homotopy equivalence, from which we get a homotopy equivalence $X_1' \to X$ by identifying back the two copies of $D_1'$.

c) Finally, if $h$ is a homotopy equivalence then $\theta(h)$ is zero, because we can assume $W$ to be a disc from the beginning, by the fibering theorem [20], so that $\hat{X}$ is a disc and $\theta(\hat{h}) = 0$.

This shows in particular that, in the case we are interested in, $\rho'(Q^{4k}) = \theta(h) \in \mathbb{Z}_2$ is the obstruction to finding an embedded $Q^{4k-2} \subset Q^{4k}$ with complement homotopy equivalent to $X$, that is, the obstruction to finding an unknotted invariant sphere for $Q^{4k}$, which is the same as $\rho(Q^{4k})$. So we have proved

**Lemma 1.** $\rho(Q^{4k}) = \rho'(Q^{4k})$.

Consider now an embedding $Q^{4k-2} \subset Q^{4k}$, and let $X'$ be $Q^{4k} - U$, $U$ a tubular neighborhood of $Q^{4k-2}$, so, we have again a normal map $h$: $X' \to X$. We will compute $\theta(h)$ and show essentially that it can be non-zero. To compute $\theta(h)$ by the description given above, we would have to prepare $X'$ by doing surgery on $W = h^{-1}(D)$ to obtain a disc, but that would change the essential property of $X'$, that of being the complement of a $Q^{4k-2}$ in a $Q^{4k}$, in particular the property of having no homology in dimensions greater than 1. We now show how to compute $\theta(h)$ without preparing $X'$.

Let, then, in general, $h$: $X' \to X$ be a normal map, $W = h^{-1}(D)$ and let $H$: $Y \to X$ be a normal cobordism between $h$ and $h_1$: $X_1' \to X$ such that $h_1^{-1}(D) = D'$, a disc. Then by cutting along $H^{-1}(D) = V$, we again obtain a cobordism $Y$, rel. boundary, between $\hat{X}_1'$ and $V \cup \hat{X}' \cup V$, and we have

$$8\sigma(\hat{X}_1') = I(\hat{X}_1') = I(V \cup \hat{X}' \cup V) = 2I(V) + I(\hat{X}').$$

Here $I(M)$ denotes the index of the manifold with boundary $M$, and we have used Novikov's lemma. (See V.8 and take $T = 1$.) The use of Novikov's lemma in this situation can be justified by attaching $D^{4k-1} \times I$

to all manifolds, which does not change the middle dimensional homology, and then we are in the situation where we glue manifolds along boundary components.

Define $\mu(W)=I(V)$ mod. 8. $\mu$ is well defined, because if $V'$ is another normal cobordism between $W$ and a disc, rel. boundary, $V\cup(-V')$ is a framed cobordism rel. boundary between two discs, so again by the same extension of Novikov's lemma, $I(V)-I(V')=I(V\cup(-V'))$ is a multiple of 8. Therefore we have the formula

$$\theta(h)=\frac{I(\hat{X}')+2\mu(W)}{8} \quad \text{mod. } 2.$$

In the proof of Theorem 1 we will construct an invariant $S^{4k-2}$ for a $(T, \Sigma^{4k})$ by constructing its complement. The simplest type of non-trivial complement is one that fibers over $S^1$. Keeping the notation of the preceding paragraphs, we compute $\theta(h)$ in this case.

**Lemma 2.** *If $X'$ fibers over $S^1$ and $W$ is a fiber, then $I(\hat{X}')=0$. Therefore $\mu(W)$ is a multiple of 4, and $\theta(h)=0$ if and only $\mu(W)$ is a multiple of 8.*

*Proof.* Clearly $\hat{X}'=W\times I$ and $I(\hat{X}')=0$, because all intersection numbers in $W\times I$ are 0. The rest of the lemma follows from the formula proved above.

Therefore, to construct a candidate for the complement of $Q^{4k-2}$ in a $Q^{4k}$ with $\rho(Q^{4k})=1$, we must find a $\pi$-manifold $W$ with boundary a sphere, such that $\mu(W)=4$ and which admits an orientation reversing diffeomorphism. Then we should check that the mapping torus of this diffeomorphism satisfies the necessary differentiable and homology properties, so that we can glue it to a tubular neighborhood of $Q^{4k-2}$ to obtain $Q^{4k}$. All these details will be taken care of by the next lemma.

We first have to recall briefly the process of plumbing. (See [18, 35].) Given a weighted graph, we associate to it a $4k$-manifold with boundary, whose intersection form is specified by the graph. The construction of this manifold can be described as follows: if the graph has vertices $v_i$, $i=1, \ldots, s$, weighted by integers $m_i$, $i=1, \ldots, s$, one takes for each $v_i$ a copy of the total space of the $D^{2k}$ bundle over $S^{2k}$ with Euler class $m_i$ (this bundle exists for $m_i$ odd only if $k=1,2,4$: Hopf invariant 1), and one "plumbs" the $i$-th and the $j$-th bundle if and only if $v_i$ and $v_j$ are joined by an edge in the graph. This plumbing is done by identifying a product neighborhood $D^{2k}\times D^{2k}$ of the $i$-th bundle, with a similar neighborhood of the $j$-th bundle, by interchanging the fibre and the base factors. The following facts are easy consequences of the definition.

i) The manifold is paralellizable if, and only if, all the numbers $m_i$ are even.

ii) The manifold and its boundary are simply connected if, and only if, the graph is a tree ($k>1$).

iii) If the graph is a tree, the homology groups of the manifold are trivial except for the one in dimension $2k$ which is free on $s$ generators $x_i$, represented by the 0-sections of the corresponding bundles.

iv) The intersection form of the manifold is given by $x_i \cdot x_i = m_i$, and $x_i \cdot x_j = 1$ or 0, according to whether $v_i$ and $v_j$ are joined by an edge in the graph or not, if $i \neq j$.

v) If, further, the determinant of this intersection form is $d \neq 0$, the homology groups of the boundary are $\mathbb{Z}$ in dimension $4k-1$, a finite group of order $d$ in dimension $2k-1$, and trivial in all other dimensions.

We will be interested in the following two weighted graphs:

$A_4$:

$E_8$:
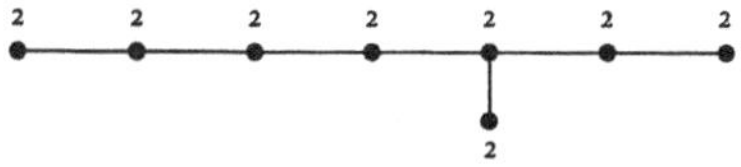


The manifold obtained by plumbing according to $A_4$ will be denoted by $A_4$ again, and its boundary by $L_5$. Let also $W = \overline{L_5 - D^{4k-1}}$. The intersection form of $A_4$ has as matrix

$$\begin{pmatrix} 2 & 1 & 0 & 0 \\ 1 & 2 & 1 & 0 \\ 0 & 1 & 2 & 1 \\ 0 & 0 & 1 & 2 \end{pmatrix}$$

which has index 4 and determinant 5. Therefore $H_{2k-1}(W) = \mathbb{Z}_5$ and all its other homology groups are trivial, and $\mu(W) = 4$.

The corresponding manifold for $E_8$ is the Milnor manifold $M_0^{4k}$, whose boundary is $\Sigma_0$. The intersection form of $M_0$ has as matrix the $E_8$-matrix

$$\begin{pmatrix} 2 & 1 & 0 & 0 & 0 & 0 & 0 & 0 \\ 1 & 2 & 1 & 0 & 0 & 0 & 0 & 0 \\ 0 & 1 & 2 & 1 & 0 & 0 & 0 & 0 \\ 0 & 0 & 1 & 2 & 1 & 0 & 0 & 0 \\ 0 & 0 & 0 & 1 & 2 & 1 & 0 & 1 \\ 0 & 0 & 0 & 0 & 1 & 2 & 1 & 0 \\ 0 & 0 & 0 & 0 & 0 & 1 & 2 & 0 \\ 0 & 0 & 0 & 0 & 1 & 0 & 0 & 2 \end{pmatrix}$$

**Lemma 3.** *$L_5 \# \Sigma_0$ is diffeomorphic to $-L_5$. In other words, $W$ admits an orientation-reversing diffeomorphism whose restriction to the boundary represents $\Sigma_0$.*

*Proof.* The proof consists in representing $M_0$ as the union of two copies of $A_4$ along $W$. Symbolically, we can view this process as that of "cracking" $E_8$ by breaking the link between its 4th and 5th vertices.

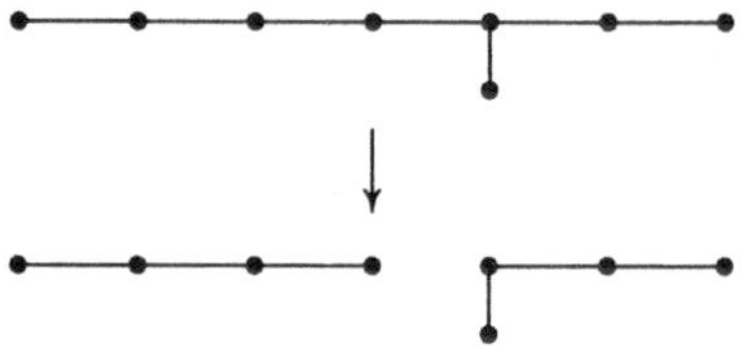

More precisely, let $e_1, \ldots, e_8$ be the basis for $H_{2k}(M_0)$, with respect to which the intersection form has as matrix $E_8$. Consider the elements $e_i'$ given by

$$e_i' = e_i \qquad i \neq 5$$

$$e_5' = -e_1 + 2e_2 - 3e_3 + 4e_4 - 5e_5 + 4e_6 - 2e_7 + 3e_8.$$

The subgroup generated by the $e_i'$ is a subgroup of $H_{2k}(M_0)$ of index 5, and the matrix of intersection numbers $e_i' \cdot e_j'$ is

$$\begin{pmatrix} 2 & 1 & 0 & 0 & 0 & 0 & 0 & 0 \\ 1 & 2 & 1 & 0 & 0 & 0 & 0 & 0 \\ 0 & 1 & 2 & 1 & 0 & 0 & 0 & 0 \\ 0 & 0 & 1 & 2 & 0 & 0 & 0 & 0 \\ 0 & 0 & 0 & 0 & 2 & 1 & 0 & 1 \\ 0 & 0 & 0 & 0 & 1 & 2 & 1 & 0 \\ 0 & 0 & 0 & 0 & 0 & 1 & 2 & 0 \\ 0 & 0 & 0 & 0 & 1 & 0 & 0 & 2 \end{pmatrix}$$

which is clearly equivalent to the direct sum of two copies of the matrix of $A_4$.

Represent $e_i'$ by embedded spheres, whose only intersections are those given by the intersection numbers. Then it is easy to see that a regular neighborhood of the union of the spheres representing $e_1', \ldots, e_4'$ is diffeomorphic to $A_4$, and so is the regular neighborhood of the union of the spheres representing $e_5', \ldots, e_8'$. We can assume these two neighborhoods to be disjoint and denote them by $A_4$ and $A'_4$ respectively.

We can further suppose that they are contained in the interior of $M_0$, except for the presence of a small tube that attaches $A_4$ to $\partial M_0$

Fig. 15

Let $K=\overline{M_0-A_4}$. From the Mayer-Vietoris sequence of $(M_0;\, A_4, K)$ we have

$$0\to H_{2k}(A_4)\oplus H_{2k}(K)\to H_{2k}(M_0)\to H_{2k-1}(W)\to 0$$

(since $H_{2k-1}(K)\approx H^{2k+1}(K, L_5)\approx H^{2k+1}(M_0, A_4)=0$) so we can identify $H_{2k}(A_4)\oplus H_{2k}(K)$ with a subgroup of $H_{2k}(M_0)$ of index 5. Since $H_{2k}(A_4)\oplus H_{2k}(A'_4)$ can be identified with the subgroup of $H_{2k}(M_0)$ generated by $\{e'_i\}$, which has also index 5, it follows that the inclusion induces an isomorphism $H_{2k}(A'_4)\to H_{2k}(K)$. It is clear now that $\overline{K-A'_4}$ is a simply connected $h$-cobordism between $L_5\#\Sigma_0$ and $-L^5$. The rest of the lemma follows trivially by removing a disc.

*Proof of Theorem* 1 *for $n=4k$ in the p.l. case:* Let $\rho(T, \Sigma^{4k})=1$ (the result follows from Theorem 2 if $\rho=0$), $Q=\Sigma/T$ and $Q_1^{4k-2}$ represent the restriction of the normal invariant of $Q$ to $P^{4k-2}$. To prove the result, it is clearly necessary and sufficient to embed $Q_1$ in $Q$, and we will do this by reconstructing $Q$ from and around $Q_1$. Let $\eta$ be the standard line bundle over $Q_1$. It was shown in the proof of Theorem 2 that $\dot{E}(\eta\oplus\eta)$ fibers over $S^1$ with fiber $\tilde{Q}_1=S^{4k-2}$, so it is p.l. homeomorphic to the mapping torus of an orientation reversing p.l. homeomorphism, which we can assume is the antipodal map, since it is in any case p.l. isotopic to it. Let $f\colon W\to W$ be the diffeomorphism constructed in Lemma 3 and let $X'$ be the mapping torus of $f$. As before $\partial X'$ is p.l. homeomorphic to the mapping torus of the antipodal map of $S^{4k-2}$, and we can form $Q'=\bar{E}(\eta\oplus\eta)\cup X'$, along their boundary.

**Lemma 4.** *$Q'$ is p.l. homeomorphic to $Q$.*

*Proof.* We first have to show that $Q'$ is a homotopy projective space. By Van Kampen, $\pi_1(Q')\approx\mathbb{Z}_2$ so we can form $\tilde{Q}'$. It is clear that $\tilde{Q}'=S^{4k-2}\times D^2\cup\tilde{X}'$, where $\tilde{X}'$ is the mapping cylinder of $f^2\colon W\to W$. To show that $\tilde{Q}'$ is a homotopy sphere it is enough to show that $\tilde{X}'$ has no homology in dimensions greater than one. For that we have to compute the homomorphism $f_*^2: H_{2k-1}(W) \to H_{2k-1}(W)$. Let $x$ generate $H_{2k-1}(W) = \mathbb{Z}_5$ and let $f_* x = kx$. If $y$ is the generator of $H^{2k}(W) \approx \operatorname{Ext}(\mathbb{Z}_5, \mathbb{Z}) = \mathbb{Z}_5$ dual to $x$, we also have $f^*(y) = ky$. Therefore

$$\begin{aligned} x &= [M] \cap y = -f_*[M] \cap y = -f_*([M] \cap f^* y) \\ &= -f_*([M] \cap k y) = -f_*(kx) = -k^2 x \end{aligned}$$

and $k^2 = -1$, so $f_*^2(x) = -x$. (Alternatively, one can use linking numbers to show this.)

Viewing $\tilde{X}'$ as the union of two copies of $W \times I$, joined along one boundary component by the identity and along the other one by $f^2$, we obtain a Mayer-Vietoris sequence, whose non-trivial part is

$$\begin{aligned} &0 \to H_{2k}(\tilde{X}') \to H_{2k-1}(W \times \dot{I}) \to H_{2k-1}(W \times I) \oplus H_{2k-1}(W \times I) \\ &\to H_{2k-1}(\tilde{X}') \to 0 \end{aligned}$$

and identifying both middle groups with $\mathbb{Z}_5 \oplus \mathbb{Z}_5$ we see that the central maps has as matrix

$$\begin{pmatrix} 1 & 1 \\ f_*^2 & 1 \end{pmatrix} = \begin{pmatrix} 1 & 1 \\ -1 & 1 \end{pmatrix}$$

and is an isomorphism. Therefore $H_i(\tilde{X}') = 0$ for $i > 1$ and $\tilde{Q}'$ is a homotopy sphere.

Having shown that $Q'$ is a homotopy projective space we now identify it with $Q$, using the classification Theorem IV.3.4. Since their normal invariants both restrict to that of $Q_1$, we have $\phi_i(Q) = \phi_i(Q')$ for $i \leqq 4k-2$. Therefore there are only two possibilities for $Q'$ according to the value of $\phi_{4k}$, and these two possibilities are clearly $Q' = Q$ or $Q' = \Sigma^2 Q_1$. But, by Theorem 2, $Q'$ is not a double suspension since

$$\rho(Q') = \rho'(Q') = \frac{\mu(W)}{4} = 1.$$

Therefore $Q = Q' \supset Q_1$, and $(T, \Sigma)$ contains the invariant sphere $\tilde{Q}_1$, which is what was to be proved.

*Remark.* Another way of showing that $\phi_{4k}$ and $\rho$ give the same information is to notice that they both change when we add a Milnor manifold. It can be conjectured that $e=\phi_{4k}$ (it is so in the trivial case), but we can think of no application of this result.

*Proof of Theorem* 1 *for* $n=4k$ *in the smooth case.* We proceed exactly as in the p.l. case, but we have to show that we can glue $\bar{E}(\eta\oplus\eta)$ and $X'$ smoothly and we have to do it carefully to get something (almost) diffeomorphic to $Q$.

Let $Q_2^{4k-1}$ be a desuspension of $Q$. Since $1=\rho(Q)=\sigma(Q'_2)$ mod. 2, and $\tilde{Q}_2=S^{4k+3}$, we have $\kappa(Q_2)=1$, so we can find $Q'_2$ with the same normal invariant as $Q'_2$ such that $\tilde{Q}'_2=\Sigma_0$ and $\sigma(Q'_2)=0$ (V.3). Then $Q'_2=\bar{E}(\eta)\cup D^{4k-1}$ and we use this diffeomorphism to identify $\dot{E}(\eta)$ with $S^{4k-2}$. If we think of $\dot{E}(\eta)$ as the fibre of $\dot{E}(\eta\oplus\eta)\to S^1$, then we can view $\dot{E}(\eta\oplus\eta)$ as the mapping torus of an orientation-reversing diffeomorphism of $S^{4k-2}$ which we proceed to identify. If we attach a handle $D^{4k-1}\times I$ to $\bar{E}(\eta\oplus\eta)$ along $\dot{E}(\eta)\times I$ we get the total space of the (closed) standard line bundle over $Q'_2$, whose boundary is $\tilde{Q}_2=\Sigma_0$. But, on the other hand, this boundary can be described as the union of two copies of $D^{4k-1}$, joined along the boundary by the given diffeomorphism of $S^{4k-2}$. This means that there is a fiber-preserving diffeomorphism between $\dot{E}(\eta\oplus\eta)$ and $\partial X'$. Let $Q'=\bar{E}(\eta\oplus\eta)\cup X'$, using this diffeomorphism.

**Lemma 4′.** *$Q'$ is diffeomorphic to a suspension of $Q_2\#\Sigma$, for some $\Sigma\in\theta^{4k-1}(\partial\pi)$ such that $2\Sigma=0$.*

*Proof.* $\bar{E}(\eta)\cup W$ is a characteristic submanifold for $Q$, which is clearly normally cobordant to $\bar{E}(\eta)\cup D^{4k-1}=Q'_2$. Therefore, the normal invariants of $Q'$ and $Q$ agree when restricted to $P^{4k-1}$. Let $Q''_2$ be a desuspension of $Q'$ with $\sigma(Q''_2)=\sigma(Q_2)$. This can clearly be accomplished since, as before, $\sigma(Q''_2)\equiv\rho(Q')=\rho(Q)\equiv\sigma(Q_2)$ mod. 2. Since $Q''_2$ and $Q_2$ have also the same normal invariant, by Theorem 2′, IV.3.3, we must have $Q''_2=Q_2\#\Sigma_1$, for $\Sigma_1\in\theta^{4k-1}(\partial\pi)$, but since $S^{4k-1}=\tilde{Q}''_2=\tilde{Q}_2\#2\Sigma_1$, the result is proved.

Now Lemma 4′ implies, in particular, that $Q$ and $Q'$ become diffeomorphic when we remove a tubular neighborhood of $S^1=P^1$ from each. Since $Q_1\subset Q'$ can be moved away from $S^1$, it follows that $Q_1\subset Q$, and Theorem 1 is finally proved.

## VI.4 Cobordism Classes of Invariant Knots

In this section we begin the study of equivariant concordance classes of invariant codimension 2 spheres. In order to underline the analogy with standard knot theory, we will refer to an invariant $\Sigma^{n-2}$ for $(T,\Sigma^n)$ as an *invariant knot*, and we say that two invariant knots $\Sigma_1$ and $\Sigma_2$ for

$(T, \Sigma^n)$ are *equivariantly cobordant* if they are equivariantly concordant, i.e. if there is an $h$-cobordism $V \subset \Sigma^n \times I$ between $\Sigma_1$ and $\Sigma_2$ which is invariant under $T \times 1$.

The existence problem for invariant knots having been solved in the preceding section, two problems naturally arise:

1) Which knot cobordism classes can be realized as invariant knots of a given involution?

2) Classify invariant knots up to equivariant cobordism.

We have only fragmentary results regarding these questions, which we will include for completeness, without proof but with some comments. We refer the reader to Levine's paper [49] for the definition of knot cobordism groups and the associated invariants, such as Seifert matrix, index, and Alexander polynomial.

We have shown that an involution $(T, \Sigma^{4k+3})$ admits a trivial invariant knot if, and only if, $\sigma(T, \Sigma^{4k+3})=0$. This suggests that there could be a general relation between $\sigma(T, \Sigma^{4k+3})$ and an invariant of the knot type of an invariant knot for $(T, \Sigma^{4k+3})$. But this invariant cannot be strictly a cobordism invariant, because we can add equivariantly to an invariant knot two copies of any other knot $K$, thus changing the cobordism class of the invariant knot by $2K$. This observation, together with our next theorem, shows that all we can hope for is a mod. 2 relation between $\sigma$ and a cobordism invariant of an invariant knot. We have not found this relation so far, but the examples given below give some hope that the Alexander polynomial of the knot might do the trick.

**Theorem 1.** a) *Every involution* $(T, \Sigma^{4k+3})$, $k>0$ *admits an invariant knot which has index* 0.

b) *If* $\sigma(T_1, \Sigma_1^{4k+3}) \equiv \sigma(T_2, \Sigma_2^{4k+3})$ *mod.* 2, $k>0$, *then* $(T_1, \Sigma_1)$ *and* $(T_2, \Sigma_2)$ *admit invariant knots* $K_1$ *and* $K_2$ *which are cobordant.*

c) *If* $\sigma(T, \Sigma^{4k+3})$ *is even,* $k>0$, *then* $(T, \Sigma)$ *admits an invariant knot which is null-cobordant.*

The proof, whose details can be found in [55], is mostly computational and is based in the direct construction given in II.4 and II.7. Starting from any involution $(T_0, \Sigma_0)$ with $\sigma(T_0, \Sigma_0)=0$, one can add handles equivariantly to an invariant $S^{4k+2}$ outside an invariant $S=S^{4k+1} \subset S^{4k+2}$, obtaining a characteristic submanifold $W=V \cup_S TV$. By choosing a new admissible set for $H_{2k+1}(W)$, one can construct a new involution with characteristic submanifold $W$, with any value of $\sigma$, for which $S \subset W$ is an invariant knot which bounds the manifold $V$. The Seifert matrix of this knot can be computed directly by expressing the basis of $H_{2k+1}(V)$ in terms of the admissible set of $H_{2k+1}(W)$ used in the construction. Using in particular the matrices $P$, $Q$ given in II.4, one obtains an involution $(T_1, \Sigma_1)$ with $\sigma(T_1, \Sigma_1)=1$ and an invariant knot with Seifert matrix

$$A=\begin{pmatrix} 4 & 3 & 3 & 1 & -4 & -3 & -6 & -3 \\ 3 & 0 & -3 & -1 & -1 & 1 & 2 & 0 \\ 3 & -3 & -8 & -2 & 1 & 6 & 8 & 3 \\ 1 & -1 & -2 & 0 & 0 & 2 & 2 & 0 \\ -3 & -1 & 1 & 0 & 2 & 0 & 1 & 1 \\ -3 & 2 & 6 & 2 & 0 & -4 & -6 & -2 \\ -6 & 2 & 9 & 2 & 1 & -6 & -8 & -2 \\ -3 & 0 & 3 & 1 & 1 & -2 & -2 & 0 \end{pmatrix}.$$

By a repeated use of these matrices one finds an involution $(T_s, \Sigma_s)$ with $\sigma(T_s, \Sigma_s)=s$ and with an invariant knot with Seifert matrix $sA$, the direct sum of $s$ copies of $A$. (The case $s<0$ can be dealt with by a change of orientation.) Computing the index of $A+A^t$ to be 0, a) follows. By adding equivariantly a number of copies of a knot with Seifert matrix $-A$ we get an invariant knot whose Seifert matrix is cobordant to 0 or to $A$, according to whether $s$ is even or odd. This proves b) and c).

The Alexander polynomial of the matrix $A$ is

$$\Delta_A(t)=4t^8-9t^7+10t^6-11t^5+13t^4-11t^3+10t^2-9t+4$$

which can be shown to be irreducible. This implies that $A$ is not cobordant to the double of any Seifert matrix. Thus we have the

**Corollary.** a) *If $\sigma(T, \Sigma^{4k+3})$ is even, $k>0$, then for every knot $K$, $(T, \Sigma)$ admits an invariant knot cobordant to $2K$.*

b) *If $\sigma(T, \Sigma^{4k+3})$ is odd, $k>0$, then $(T, \Sigma)$ admits an invariant knot which is not cobordant to $2K$ for any knot $K$.*

Now we turn to the problem of determining the equivariant cobordism classes of invariant knots. Recall that a knot $\Sigma^n \subset \Sigma^{n+2}$ is called simple if the homotopy groups of the complement $\Sigma^{n+2}-\Sigma^n$ coincide with those of $S^1$ in dimensions $\leqq \left[\frac{n-1}{2}\right]$.

**Theorem 2.** *Every invariant knot is equivariantly cobordant to a simple knot.*

This result corresponds to the preparatory lemma used in computing the knot cobordism groups ([49], Lemma 4). The proof of this lemma cannot be carried to the invariant case because it uses an engulfing result that has no equivariant counterpart. Theorem 2 can be proved by doing equivariant surgery on the complement of an invariant tubular neighborhood of a knot, rel. boundary, in such a way that both the result and the trace of these surgeries have the homology groups of $S^1$ in all dimensions, and the homotopy groups of $S^1$ below the middle dimension. Then the trace of these surgeries can be glued equivariantly to $S^n \times D^2 \times I$ along $S^n \times S^1 \times I$, obtaining an $h$-cobordism which can be identified with $(T, \Sigma^{n+2}) \times I$ inside which $S^n \times I$ is an equivariant cobordism between the original knot and a simple one.

This is all we can say for the moment. The next step would be to identify the obstruction to killing the middle dimensional homotopy groups by a sequence of surgeries whose traces have the homology groups of $S^1$. The problem is further complicated by the fact that we can have two invariant trivial knots which are not equivariantly cobordant. Cf. Corollary VI.3 and the paragraph preceding it, where the equivariant cobordism classes of invariant trivial knots are computed.

*Added in proof.* Further results on codimension 2 problems can be found in the author's Invariant Knots and Surgery in Codimension 2, Proceedings of the I.C.M., Nice 1970.

# Some Unsolved Problems

1. The compatibility relation between matrices (II.4) probably deserves a more detailed study. Given $H$ and $H'$, what are all the possible ways of choosing $P$ and $Q$ to give the compatibility? This question seems related to the definition of Wall's odd dimensional surgery obstructions [82].

2. Can one say something more specific about the involutions of homology 3-spheres (II.5)? Can the irreducible ones be classified, or at least their fundamental groups? Notice that in the simplest case with $\sigma \neq 0$ the fundamental group is presented by 16 generators and 16 relations. In the (unlikely) case that one of these turned out to be the trivial group, a counterexample to the Poincaré conjecture would have been found. A careful study of these involutions would involve the study of question 1 and the description of all possible liftings of an automorphism of $H_1(W)$ to an automorphism of $\pi_1(W)$.

3. Is the map $\mathbb{Z}_2 \to hS(P^{4k+1})$ injective in the dimensions where the Kervaire invariant conjecture holds? In other words, is $Q^{4k+1} \# \Sigma_0^{4k+1}$ different from $Q^{4k+1}$, where $\Sigma_0$ is the Kervaire sphere, provided that $\Sigma_0 \neq S^{4k+1}$?

Similarly, does $\theta^{4k+3}(\partial\pi)$ always act freely on $hS(P^{4k+3})$, even in the cases where Browder's theorem (V.7) is not known to apply?

4. In Corollary 2, II.3, are there exactly 2 isotopy classes of embeddings of $P^{4k+1}$ in $P^{4k+2}$? In other words, does concordance (pseudo-isotopy) imply isotopy for embeddings in this situation? The work of J. Cerf on pseudo-isotopy of diffeomorphisms might be useful for this question. Can something be said about the case of Corollary VI.3?

5. Are there curious involutions (V.3) in all odd dimensions? Except for those in dimension 7, and some trivial ones where $\Sigma_0 = S^{4k+1}$, no examples are known, and all known methods of constructing new involutions out of old do not change the invariant $\kappa$. Perhaps there are simple examples in dimension 15, similar to the Hirsch-Milnor involution.

Can the invariant $\kappa$ be extended naturally to the cases where $\Sigma^n \notin \theta^n(\partial\pi)$? Is it a p.l. invariant? These questions are related to the realization of the invariant $\rho$ in the smooth case (VI.3).

6. The operation of taking an odd multiple of an involution, or, more generally, of adding together an odd number of involutions is not yet fully understood (V.5.2). What does it mean in terms of mappings into $G/0$? Does it have any relation to the Livesay-Thomas method of adding involutions ([51])?

7. Compute $\sigma(T, \Sigma_a)$, VI.2.2.

8. Is there an even dimensional exotic sphere that admits an involution? This could be checked using the methods of V.6.2 and the recent computations of homotopy groups of spheres.

9. Does every $(T, \Sigma^{4k+3})$ admit an invariant null-cobordant knot? A positive answer would give another proof of Theorem 1, VI.3 for $n=4k+4$, since then the null-cobordant knot for a desuspension of $(T, \Sigma^{4k+4})$ bounds a disc $D$ in one of the components of the complement of this desuspension, and $D \cup TD$ is an invariant knot for $(T, \Sigma^{4k+4})$. A negative answer could be obtained by identifying $\sigma$ (mod. 2) with a cobordism invariant of an invariant knot (VI.4).

10. Compute the cobordism classes and equivariant cobordism classes of invariants knots for a given $(T, \Sigma^n)$ (VI.4). In particular, are any two invariant knots for $(T, \Sigma^{2k})$ equivariantly cobordant?

# References


1. Adams, J.F.: On the groups $J(X)$ – IV. Topology **5**, 21–71 (1966).
2. Arf, C.: Untersuchungen über quadratische Formen in Körpern der Charakteristic 2. J. Reine Angew. Math. **183**, 148–167 (1941).
3. Atiyah, M.F., Bott, R.: A Lefschetz fixed point formula for elliptic complexes. II. Applications. Ann. of Math. **88**, 451–491 (1968).
4. — Singer, I.M.: The index of elliptic operators. III. Ann. of Math. **87**, 546–604 (1968).
5. Berstein, I.: Involutions with non-zero Arf invariant. Bull. Amer. Math. Soc. **74**, 678–682 (1968).
6. — Livesay, G.R.: Non-unique desuspension of involutions. Inventiones Math. **6**, 56–66 (1968).
7. Boardman, J.M., Vogt, R.M.: Homotopy-everything $h$-spaces. Bull. Amer. Math. Soc. **74**, 1117–1124 (1968).
8. Borel, A.: Seminar on transformation groups. Ann. of Math. Studies, No. 46, Princeton, 1960.
9. Bredon, G.: A $\pi_*$-module structure for $\theta_*$ and applications to transformation groups. Ann. of Math. **86**, 434–448 (1967).
10. Brieskorn, E.: Beispiele zur Differentialtopologie von Singularitäten. Invent. Math. **2**, 1–14 (1966).
11. Browder, W.: Homotopy type of differentiable manifolds. Colloq. on Alg. Top., Aarhus (1962), p. 42–46.
12. — Embedding 1-connected manifolds. Bull. Amer. Math. Soc. **72**, 225–231 (1966).
13. — On the action of $\theta^n(\partial\pi)$. Differential and Combinatorial Topology, a symposium in honor of Marston Morse. Princeton, N.J.: Princeton Univ. Press 1965.
14. — Diffeomorphisms of 1-connected manifolds. Trans. Amer. Math. Soc. **128**, 155–163 (1967).
15. — Embedding smooth manifolds. Proc. I.C.M. Moskow (1966).
16. — The Kervaire invariant of framed manifolds and its generalization. Ann. of Math. **90**, 157–186 (1969).
17. — Cobordism invariants, the Kervaire invariant and fixed point free involutions (to appear).
18. — Surgery on simply connected manifolds (to appear).
19. — Hirsch, M.W.: Surgery on piecewise linear manifolds and applications. Bull. Amer. Math. Soc. **72**, 959–964 (1966).
20. — Levine, J.: Fibering manifolds over a circle. Comment. Math. Helv. **40**, 153–160 (1966).
21. — Livesay, G.R.: Fixed point free involutions on homotopy spheres. Bull. Amer. Math. Soc. **73**, 242–245 (1967).
22. — — (to appear).
23. — Petrie, T.: Semi-free actions and diffeomorphisms (to appear).
24. Conner, P.E., Floyd, E.E.: Differentiable periodic maps. Ergebnisse der Mathematik und ihrer Grenzgebiete 33. Berlin-Göttingen-Heidelberg: Springer 1964.
25. Giffen, C.H.: Desuspendability of free involutions on Brieskorn spheres. Bull. Amer. Math. Soc. **75**, 426–429 (1969).

26. Giffen, C.H.: Smooth homotopy projective spaces. Bull. Amer. Math. Soc. **75**, 509–513 (1969).
27. — Weakly complex involutions and cobordism of projective spaces. Ann. of Math. **90**, 418–432 (1969).
28. Haefliger, A.: Plongements différentiables de variétés dans variétés. Comment. Math. Helv. **36**, 47–82 (1961).
29. — Knotted spheres and related geometric problems. Proc. I.C.M. Moskow (1966).
30. Higman, G.: The units of group rings. Proc. London Math. Soc., Ser. II, **46**, 231–248 (1940).
31. Hirsch, M.W., Milnor, J.: Some curious involutions of spheres. Bull. Amer. Math. Soc. **70**, 372–377 (1964).
32. Hirzebruch, F.: Singularities and exotic spheres. Seminaire Bourbaki **1966/67**, No. 314.
33. — Involutionen auf Mannigfaltigkeiten. Proceedings of the Conference on Transformation Groups, New Orleans, 1967, p. 148–166. Berlin-Heidelberg-New York: Springer 1968.
34. — Jänich, K.: Involutions and singularities. Proc. Int. Colloq. on Algebraic Geometry, Bombay 1968 (to appear).
35. — Mayer, K.H.: $0(n)$-Mannigfaltigkeiten, exotische Sphären und Singularitäten. Lecture Notes in Mathematics. 57. Berlin-Heidelberg-New York: Springer 1968.
36. Jänich, K., Ossa, E.: On the signature of an involution. Topology **8**, 27–30 (1969).
37. Jones, B.W.: The arithmetic theory of quadratic forms. Carus Mathematical Monographs 10. The Math. Ass. of America, 1950.
38. Karrass, A., Magnus, W., Solitar, D.: Combinatorial group theory. New York: Interscience Publishers 1966.
39. Kervaire, M.: A manifold which does not admit any differentiable structure. Comment. Math. Helv. **34**, 257–270 (1960).
40. — Le théorème de Barden-Mazur-Stallings. Comment. Math. Helv. **40**, 31–42 (1965).
41. — On higher dimensional knots. Differential and Combinatorial Topology, a symposium in honor of Marston Morse. Princeton, N.J.: Princeton University Press 1965.
42. — Milnor, J.: Groups of homotopy spheres I. Ann. of Math. **77**, 504–537 (1963).
43. — — On 2-spheres in 4-manifolds. Proc. Nat. Acad. Sci. U.S.A. **47**, 1651–1657 (1961).
44. Kirby, R.C.: Stable homeomorphisms and the annulus conjecture. Ann. of Math. **89**, 575–582 (1969).
45. — Siebenmann, L.C.: On the triangulation of manifolds and the Hauptvermutung. Bull. Amer. Math. Soc. **75**, 742–749 (1969).
46. — — For manifolds the Hauptvermutung and the triangulation conjecture are false. Notices Amer. Math. Soc. **16**, 695 (1969).
47. Lee, R.: Non-existence of free differentiable actions of $S^1$ and $\mathbf{Z}_2$ on homotopy spheres. Proceedings of the Conference on Transformation Groups, New Orleans 1967, p. 208–209. Berlin-Heidelberg-New York: Springer 1968.
48. Levine, J.: Unknotting spheres in codimension two. Topology **4**, 9–16 (1965).
49. — Knot Cobordism groups in codimension two. Comment. Math. Helv. **44**, 229–244 (1969).
50. Livesay, G.R.: Fixed point free involutions on the 3-spheres. Ann. of Math. **72**, 603–611 (1960).
51. — Thomas, C.B.: Involutions on homotopy spheres. Proceedings of the Conference on Transformation Groups, New Orleans, 1967, p. 143–147. Berlin-Heidelberg-New York: Springer 1968.
52. López de Medrano, S.: Involutions of homotopy spheres and homology 3-spheres. Bull. Amer. Math. Soc. **73**, 727–731 (1967).
53. — Some results on involutions of homotopy spheres. Proceedings of the Conference on Transformation Groups, New Orleans, 1967. Berlin-Heidelberg-New York: Springer 1968.

54. López de Medrano, S.: Involutions. Princeton, Ph. D. Thesis, 1968.
55. — Nudos invariantes bajo involuciones. I. An. Inst. Mat. Univ. Nac. Autónoma México **8**, 81–90 (1969).
56. Mahowald, M. E.: On the order of the image of J. Topology **6**, 371–378 (1967).
57. Milnor, J.: On simply connected 4-manifolds. Symposium Internac. de Topología Algebráica, México (1956).
58. — A procedure for killing the homotopy groups of differentiable manifolds. AMS Symposium in Pure Mathematics III (1961).
59. — Lectures on the *h*-cobordism theorem. Notes by L. Siebenmann and J. Sondow. Princeton 1965.
60. — Remarks concerning spin manifolds. Differential and Combinatorial Topology, a symposium in honor of Marston Morse. Princeton, N.J.: Princeton Univ. Press 1965.
61. Montgomery, D., Yang, C. T.: Differentiable actions on homotopy 7-spheres. I. Trans. Amer. Math. Soc. **122**, 480–498 (1966); II. Proceedings of the Conference on Transformation Groups, New Orleans, 1967, p. 125–134. Berlin-Heidelberg-New York: Springer. See also: Free differentiable actions on homotopy spheres. Same Proceedings, p. 173–192.
62. — Zippin, L.: Topological transformation groups. New York: Interscience Publishers Inc. 1955.
63. Mostert, P. S.: Proceedings of the Conference on Transformation Groups, New Orleans, 1967. Berlin-Heidelberg-New York: Springer 1968.
64. Nielsen, J.: Untersuchungen zur Topologie der geschlossenen zweiseitigen Flächen. Acta Math. **50**, 189–358 (1924).
65. Novikov, S. P.: Homotopy equivalent smooth manifolds I [in Russian]. Izv. Akad. Nauk USSR, Ser. Mat. **28** (2), 365–374 (1964).
66. Olum, P.: Mappings of manifolds and notion of degree. Ann. of Math. **58**, 458–480 (1953).
67. Orlik, P.: On the Arf invariant of an involution. Canad. J. Math. (to appear).
68. — Rourke, C. P.: Free involutions on homotopy $(4k+3)$-spheres. Bull. Amer. Math. Soc. **74**, 949–953 (1968).
69. Puppe, D.: Homotopiemengen und ihre induzierten Abbildungen. I. Math. Z. **69**, 299–344 (1958).
70. Serre, J. P.: Formes bilinéaires symétriques entières à discriminant $\pm 1$. Séminaire Cartan, 1961/62, Exposé 14.
71. Siebenmann, L.: Finding a boundary for an open manifold. Princeton, Ph. D. Thesis (1965).
72. Smale, S.: On the structure of manifolds. Amer. J. Math. **84**, 387–399 (1962).
73. Spivak, M.: Spaces satisfying Poincaré duality. Topology **6**, 77–102 (1967).
74. Stasheff, J.: A classification theorem for fibre spaces. Topology **2**, 239–246 (1963).
75. Sullivan, D.: Triangulating homotopy equivalences. Princeton, Ph. D. Thesis (1965).
76. — Triangulating and smoothing homotopy equivalences and homeomorphisms. Geometric Topology Seminar Notes, Princeton, 1967 (mimeographed).
77. Toda, H.: Composition methods in homotopy groups of spheres. Ann. of Math. Studies 49, Princeton 1962.
78. Wall, C. T. C.: An extension of results of Novikov and Browder. Amer. J. Math. **88**, 20–32 (1966).
79. — Locally flat PL submanifolds with codimension two. Proc. Cambridge Philos. Soc. **63**, 5–8 (1967).
80. — Poincaré complexes. I. Ann. of Math. **86**, 213–245 (1967).
81. — Surgery of non-simply-connected manifolds. Ann. of Math. **84**, 217–276 (1966).

82. Wall, C.T.C.: Surgery of compact manifolds (to appear). (Benjamin.)
83. — Free piecewise linear involutions on spheres. Bull. Amer. Math. Soc. **74**, 554–558 (1968).
84. Zeeman, E.C.: Seminar on combinatorial topology, Institut des Hautes Etudes Scientifiques, 1963.

# Index

# List of Symbols



[1] The symbols $\sigma(f)$ and $\tau(f)$ can have different meanings according to whether $f$ is a normal map or an equivariant homotopy equivalence.

# Ergebnisse der Mathematik und ihrer Grenzgebiete

---

1. Bachmann: Transfinite Zahlen. DM 42,—; US $ 11.60
2. Miranda: Partial Differential Equations of Elliptic Type. DM 58,—; US $ 16.00
4. Samuel: Méthodes d'Algèbre Abstraite en Géométrie Algébrique. DM 29,—; US $ 8.00
5. Dieudonné: La Géométrie des Groupes Classiques. DM 38,—; US $ 10.50
6. Roth: Algebraic Threefolds with Special Regard to Problems of Rationality. DM 22,—; US $ 6.10
7. Ostmann: Additive Zahlentheorie. 1. Teil: Allgemeine Untersuchungen. DM 42,—; US $ 11.60
8. Wittich: Neuere Untersuchungen über eindeutige analytische Funktionen. DM 31,—; US $ 8.60
11. Ostmann: Additive Zahlentheorie. 2. Teil: Spezielle Zahlenmengen. DM 28,—; US $ 7.70
13. Segre: Some Properties of Differentiable Varieties and Transformations. DM 46,—; US $ 12.70
14. Coxeter/Moser: Generators and Relations for Discrete Groups. DM 35,—; US $ 9.70
15. Zeller/Beckmann: Theorie der Limitierungsverfahren. DM 64,—; US $ 17.60
16. Cesari: Asymptotic Behavior and Stability Problems in Ordinary Differential Equations. 3rd edition in preparation
17. Severi: Il teorema di Riemann-Roch per curve-superficie e varietà questioni collegate. DM 26,—; US $ 7.20
18. Jenkins: Univalent Functions and Conformal Mapping. DM 37,—; US $ 10.20
19. Boas/Buck: Polynomial Expansions of Analytic Functions. DM 18,—; US $ 5.00
20. Bruck: A Survey of Binary Systems. DM 46,—; US $ 12.70
21. Day: Normed Linear Spaces. DM 17,80; US $ 4.90
23. Bergmann: Integral Operators in the Theory of Linear Partial Differential Equations. DM 40,—; US $ 11.00
25. Sikorski: Boolean Algebras. DM 42,—; US $ 11.60
26. Künzi: Quasikonforme Abbildungen. DM 43,—; US $ 11.90
27. Schatten: Norm Ideals of Completely Continuous Operators. DM 26,—; US $ 7.20
28. Noshiro: Cluster Sets. DM 40,—; US $ 11.00
29. Jacobs: Neuere Methoden und Ergebnisse der Ergodentheorie. Vergriffen
30. Beckenbach/Bellman: Inequalities. DM 38,—; US $ 10.50
31. Wolfowitz: Coding Theorems of Information Theory. DM 30,—; US $ 8.30
32. Constantinescu/Cornea: Ideale Ränder Riemannscher Flächen. DM 75,—; US $ 20.70
33. Conner/Floyd: Differentiable Periodic Maps. DM 29,—; US $ 8.00
34. Mumford: Geometric Invariant Theory. DM 24,—; US $ 6.60
35. Gabriel/Zisman: Calculus of Fractions and Homotopy Theory. DM 42,—; US $ 11.60
36. Putnam: Commutation Properties of Hilbert Space Operators and Related Topics. DM 31,—; US $ 8.60
37. Neumann: Varities of Groups. DM 51,—; US $ 14.10
38. Boas: Integrability Theorems for Trigonometric Transforms. DM 20,—; US $ 5.50
39. Sz.-Nagy: Spektraldarstellung linearer Transformationen des Hilbertschen Raumes. DM 20,—; US $ 5.50
40. Seligman: Modular Lie Algebras. DM 43,—; US $ 11.90
41. Deuring: Algebren. DM 26,—; US $ 7.20
42. Schütte: Vollständige Systeme modaler und intuitionistischer Logik. DM 26,—; US $ 7.20
43. Smullyan: First-Order Logic. DM 36,—; US $ 9.90

44. Dembowski: Finite Geometries. DM 68,—; US $ 17.00
45. Linnik: Ergodic Properties of Algebraic Fields. DM 44,—; US $ 12.10
46. Krull: Idealtheorie. DM 28,—; US $ 7.70
47. Nachbin: Topology on Spaces of Holomorphic Mappings. DM 18,—; US $ 5.00
48. A. Ionescu Tulcea/C. Ionescu Tulcea: Topics in the Theory of Lifting. DM 36,—; US $ 9.90
49. Hayes/Pauc: Derivation and Martingales. DM 48,—; US $ 13.20
50. Kahane: Séries de Fourier Absolument Convergents. DM 44,—; US $ 12.10
51. Behnke/Thullen: Theorie der Funktionen mehrerer komplexer Veränderlichen. DM 48,—; US $ 13.20
52. Wilf: Finite Sections of Some Classical Inequalities. DM 28,—; US $ 7.70
53. Ramis: Sous-ensembles analytiques d'une variété banachique complexe. DM 36,—; US $ 9.90
54. Busemann: Recent Synthetic Differential Geometry. DM 32,—; US $ 8.80
55. Walter: Differential and Integral Inequalities. DM 74,—; US $ 20.40
56. Monna: Analyse non-archimédienne. DM 38,—; US $ 10.50
57. Alfsen: Compact Convex Sets and Boundary Integrals. DM 46,—; US $ 12.70
58. Greco/Salmon: Topics in m-Adic Topologies. DM 24,—; US $ 6.60
59. López de Medrano: Involutions on Manifolds. DM 36,—; US $ 9.90
60. Sakai: C*-Algebras and W*-Algebras. DM 66,—; US $ 18.20
61. Zariski: Algebraic Surfaces. In preparation